"Richter's book moved me to tears. notion that the earth was given to hu animals domestic and wild, and str Elizabeth Harvey Roberts, The Natur

"The environment is one of the most pressing as well as polarizing issues today. As Sandra Richter points out, this issue is not simply a political question but a human question; it's a matter of life and death. For that reason *Stewards of Eden* is critically important as it skillfully and clearly interprets the relevant biblical passages. All Christians should read this book."

Tremper Longman III, professor emeritus of biblical studies at Westmont College

"Sandra Richter's *Stewards of Eden* is a compelling and passionate call to action for a paralyzed church. Richter laments that young Christians hear the message that they should not advocate for creation and calls all Christians to a value system of holiness. By the end of the book one feels equipped to live out a call to care for the amazing creation our Creator loves."

Kristen Page, Ruth Kraft Strohschein Distinguished Chair and Professor of Biology at Wheaton College

"Sandra L. Richter presents a forceful and compelling prophetic assessment of the idolatries that influence many of our current economic and social practices and that deter us from properly imaging the Creator in caring for and protecting the creation. She reminds us that human flourishing is inextricably linked to the flourishing of creation as a whole."

Blaine B. Charette, professor of New Testament at Northwest University

"In *Stewards of Eden*, Sandra Richter has not only given us a biblical theology of creation care, she has profiled a biblical ethos that strikes a proper ethical balance between economic productivity and sustainability. I highly commend this work to pastors as a resource for doing biblical discipleship, equipping Christ followers to prioritize their calling as advocates for God's creation in this time when many are prioritizing short-term convenience and temporary prosperity."

Darryl Williamson, lead pastor of Living Faith Bible Fellowship, Tampa

"Creation is groaning like a woman in travail, and we are called to be midwives, delivering God's creation to future generations in as good or better shape than we received it. Sandy Richter is a gifted biblical scholar and teacher, the right messenger with the right message at the right time."

Matthew Sleeth, author of *Reforesting Faith* and executive director of Blessed Earth

"*Stewards of Eden* challenges, disturbs, and encourages us to recognize God's consistent concern for the well-being of all creation and our responsibility as God's stewards. Sandra Richter's work is both very readable and richly insightful—I highly recommend it!"

Christine D. Pohl, professor emeritus of Christian ethics at Asbury Theological Seminary

"My soul comes alive hiking mountain trails or kayaking across a saltwater bay. I remember and give thanks to the Creator. I encourage my congregation to sabbath and rest in creation. Yet, how is creation stewardship expressed in our day-to-day living as followers of Jesus? There I fall short. *Stewards of Eden* provides a compelling biblical analysis and a wake-up call to our responsibility as those who bear the imprint of the Creator. Dr. Richter's wisdom paired with practical suggestions awakens a movement toward justice for all of creation."

Melissa Maher, lead pastor of Mercy Street, Houston, Texas

"Sandra Richter is a learned and seasoned interpreter of the Scriptures. She brings into focus how care for our broken world occupies a vital place in robust biblical faith and discipleship. Bringing God's living Word into contact with his hurting world, she issues a call that every follower of Christ will need to take seriously."

Lawson Stone, professor of Old Testament at Asbury Theological Seminary

"Sandra Richter does a brilliant job of showing us in her contagious, passionate way how a theology of creation can intersect with practical guidelines for stewardship. Anyone who reads this book cannot avoid hearing the universal call to think rightly and justly about issues of sustainability for the future. The discussion questions provided with each chapter prompt reflection and evoke a responsiveness for all readers."

Nadine C. Folino-Rorem, professor of biology at Wheaton College

"This book is for much more than Jesus-loving environmentalists, it is for anyone who takes the Bible seriously. *Stewards of Eden* offers a long-overdue reminder that the story of the Bible and the work of Jesus are centrally about the restoration of the human vocation of glorifying God through the stewardship of the creation in his image."

Matt Ayars, president and dean of the School of Theology, Emmaus University

"Sandra Richter marshals evidence from her decades of first-rate engagement with biblical theology, her extraordinary knowledge of pertinent Ancient Near East sources, and her awareness of unconscionable contemporary environmental disasters in order to challenge the church. The task of the church is to nurture lives of restraint as we use resources and charity as we share them."

Elaine A. Phillips, Harold John Ockenga Distinguished Professor of Biblical and Theological Studies, Gordon College

"We can scarcely overstate the relevance of the subject of this volume for Christians who identify as evangelicals. Combining sound interpretation, engaging literary style, and awareness of the current urgency, Richter provides Christians with an invaluable resource for study, reflection, discussion, and action. This book should be required reading for all who seek to make a difference in this world."

Daniel I. Block, Gunther H. Knoedler Professor Emeritus of Old Testament, Wheaton College

"Arguing that proper care—and even love—for ecosystems and the creatures (including humans) that inhabit them should surpass our superficial partisan divides, Richter provides a deep and compelling foundation for a Christian environmental and animal welfare ethic."

Amanda M. Sparkman, associate professor of biology at Westmont College

"Richter helps relocate us in the *life* of *God's* beloved world, setting us in our proper place as life promoters, glory sharers, and human stewards of creation, whose destiny is inextricably joined with our own—ultimately so in our still-incarnate Lord. Read and prayerfully live the truth of this book as you read and hear anew God's truth."

Cherith Fee Nordling, author of *Knowing God by Name*

"A number of biblical scholars have demonstrated that 'creation care' is deeply rooted in the Scriptures. But Dr. Richter's book stands apart in its focus on examples from everyday life that illustrate and bring home to us the biblical principles she highlights."

Douglas Moo, Wessner Chair of Biblical Studies at Wheaton College and chair

STEWARDS OF EDEN

SANDRA L. RICHTER

WHAT SCRIPTURE SAYS ABOUT THE ENVIRONMENT AND WHY IT MATTERS

Academic

An imprint of InterVarsity Press
Downers Grove, Illinois

InterVarsity Press
P.O. Box 1400, Downers Grove, IL 60515-1426
ivpress.com
email@ivpress.com

©2020 by Sandra L. Richter

All rights reserved. No part of this book may be reproduced in any form without written permission from InterVarsity Press.

InterVarsity Press® is the book-publishing division of InterVarsity Christian Fellowship/USA®, a movement of students and faculty active on campus at hundreds of universities, colleges, and schools of nursing in the United States of America, and a member movement of the International Fellowship of Evangelical Students. For information about local and regional activities, visit intervarsity.org.

All Scripture quotations, unless otherwise indicated, are the author's translation.

While any stories in this book are true, some names and identifying information may have been changed to protect the privacy of individuals.

Excerpt from "A Biblical Theology of Creation Care," *Sandra Richter,* Asbury Journal 62.1 (2007): 67-76. *Used by permission of The Asbury Journal.*

Excerpt from Bulletin for Biblical Research, 20.3, *"Environmental Law in Deuteronomy: One Lens on a Biblical Theology of Creation Care", Sandra Richter (author), Richard Hess (editor), copyright © 2010. This article is used by permission of The Pennsylvania State University Press.*

Excerpt from Bulletin for Biblical Research, 24.3, *"Environmental Law: Wisdom from the Ancients", Sandra Richter (author), Richard Hess (editor), copyright © 2014. This article is used by permission of The Pennsylvania State University Press.*

Excerpt from Handbook of Religion *edited by Terry C. Muck, Harold A. Netland, and Gerald R. McDermott, copyright © 2014. Used by permission of Baker Academic, a division of Baker Publishing Group.*

Cover design and image composite: David Fassett
Interior design: Jeanna Wiggins
Images: *floral design:* © CSA Images / Getty Images
   Great Barrier Reef: © Satellite Earth Art / 500px Prime / Getty Images
   green plant leaf: © Iev Anc / EyeEm / Getty Images
   green plants: © Sven Krobot / EyeEm / Getty Images
   glacial floes-Iceland: © Abstract Aerial Art / Digital Vision / Getty Images
   green forest: © Alexander Schitschka / EyeEm / Getty Images
   cardboard texture: © Katsumi Murouchi / Moment Collection / Getty Images
   mountain view: © Ivan / Moment Collection / Getty Images
   wooden cross section: © Sergey Ryumin / Moment Collection / Getty Images

ISBN 978-0-8308-4926-0 (print)
ISBN 978-0-8308-4927-7 (digital)

Printed in the United States of America ∞

InterVarsity Press is committed to ecological stewardship and to the conservation of natural resources in all our operations. This book was printed using sustainably sourced paper.

Library of Congress Cataloging-in-Publication Data
A catalog record for this book is available from the Library of Congress.

| P | 23 | 22 | 21 | 20 | 19 | 18 | 17 | 16 | 15 | 14 | 13 | 12 | 11 | 10 | 9 | 8 | 7 | 6 | 5 | 4 | 3 |
| Y | 39 | 38 | 37 | 36 | 35 | 34 | 33 | 32 | 31 | 30 | 29 | 28 | 27 | 26 | 25 | 24 | 23 | 22 | 21 | | |

CONTENTS

ACKNOWLEDGMENTS	ix
INTRODUCTION: Can a Christian Be an Environmentalist?	1
1 Creation as God's Blueprint	5
2 The People of the Old Covenant and Their Landlord	15
3 The Domestic Creatures Entrusted to ʾādām	29
4 The Wild Creatures Entrusted to ʾādām	48
5 Environmental Terrorism	60
6 The Widow and the Orphan	67
7 The People of the New Covenant and Our Landlord	91
CONCLUSION: How Should We Then Live?	106
APPENDIX: Resources for the Responsive Christian	113
NOTES	119
BIBLIOGRAPHY	142
ART CREDITS	154
SUBJECT INDEX	155
SCRIPTURE INDEX	157

FOR MY DAUGHTERS

Noël and Elise

image bearers of their Creator
and beloved in every way

ACKNOWLEDGMENTS

THIS LITTLE BOOK has been a long time in the birthing. My passion for God's good creation and for good theology in addressing the stewardship of this good creation has accompanied me throughout my career. As a result I have spoken, taught, and written on this topic in more venues than I can recall to more audiences than I can list. As I am an academic, many of these venues have been colleges, graduate schools, and professional academic societies. But many have been popular as well. Over the years I've served on the board of Blessed Earth, attended the Lausanne Creation Care conferences, and built relationships with numerous national, parachurch and missionary organizations that are spending their lives making a difference for both the environment and the marginalized dependent on it. My dream has been to take this material, which has evolved over a decade of inquiry, and has spoken into the lives and hearts of so many (not the least my own), and place it into the hands of the everyday believer in a form that they can *use*. And as so many within the church simply don't know what to do with the topic of environmental stewardship, my ambition has been to put my research into a format that is as accessible to the college student as it is to his or her parents and grandparents. Between these covers is that effort.

As so many have journeyed with me toward this goal, I wish to acknowledge the past publishers and organizations that have so kindly allowed me a platform to develop my thoughts on this topic. Below please find a list of all those endeavors that have found their way into print. May I also use this opportunity to thank Lawson Stone, who has so generously shared his beautiful images with me. To Matthew and Nancy Sleeth, who have offered me unending support and the best of friendships, for their organization Blessed Earth, and for allowing me to blog with them for a time. My thanks to Wheaton College, which granted Kristen Page and me the opportunity to coteach a full-semester course on this important topic to some of the finest students in the country. The Institute of Biblical Research and the Evangelical Theological Society deserve recognition for welcoming plenary and sectional presentations from me on this topic when it was still considered "edgy" and allowing me to reutilize my material in this publication. To Asbury University, Biola University, Evangel University, Living Faith Bible Fellowship in Tampa, Florida, and Arise City Summit, as well as the Mississippi Conference of the United Methodist Church, who took the risk of inviting me to address this issue as an aspect of contemporary holiness to their constituencies. To Asbury Theological Seminary, which has always supported me in this quest, and to Westmont College, which has provided me the professional space to write this book, I owe my gratitude. There are many more on the list who have listened, chastened, and joined their voices as I've developed this material. It is my great hope that, in this last incarnation of all the communications that have come before, we the church will be inspired to action. And in taking our place in this current crisis, we can do what we have always been called to do: change the world.

"The Bible and American Environmental Practice: An Ancient Code Addresses a Current Crisis." In *The Bible and the American Future*, edited by Robert Jewett with Wayne L. Alloway Jr. and John G. Lacey, 108-29. Eugene, OR: Cascade, 2009.
"A Biblical Theology of Creation Care: Is Environmentalism a Christian Value?" *Asbury Journal* 62, no. 1 (2007): 67-76.
Blog posts from Blessed Earth, 2010–2012. www.blessedearth.org/.
"Environmental Law in Deuteronomy: One Lens on a Biblical Theology of Creation Care." *Bulletin for Biblical Research* 20, no. 3 (2010): 331-54.

"Environmental Law: Wisdom from the Ancients." *Bulletin for Biblical Research* 24, no. 3 (2014): 307-29.

"Environmentalism and the Evangelical: Just the Bible for Those Justly Concerned." *Westmont Magazine*, spring 2019, 38-46.

"Religion and the Environment." In *Handbook of Religion: A Christian Engagement with Traditions, Teachings, and Practices*, edited by Terry C. Muck, Harold A. Netland, and Gerald R. McDermott, 746-55. Grand Rapids: Baker Academic, 2014.

INTRODUCTION

Can a Christian Be an Environmentalist?

THE SUBJECT MATTER of this book is, in my opinion, one of the most misunderstood topics of holiness and social justice in the Christian community today. The topic is obviously important, relevant, contemporary, and compelling. It is an issue our neighbors (both locally and globally) care about deeply. As a result, this is a subject that profoundly influences the church's witness to the world. But as I have traveled, written, and spoken on this issue in Christian circles for more than a decade, I have found that the church is largely paralyzed on this topic. From college students to CEOs, seminarians to pastors, cattle ranchers to coal miners, Californians to Kentuckians—we the church are MIA on the issue of environmental stewardship.

When I was teaching Old Testament at Wheaton College, professor Kristen Page (the Ruth Kraft Strohschein Distinguished Chair of Biology) and I won a Faith and Learning grant to launch the first-ever Wheaton course designed specifically to integrate the Bible and biology in an inquiry into environmental stewardship. Our title was "Environmental Concern for the Christian: The Bible and Biology." As professors are wont to do, we opened the first class with a seemingly innocent "icebreaker": "Introduce yourself to the class, telling us your name, your major, and why you took this course." Like most teachers, I have deployed a conversation starter like this dozens of times in an array of classroom settings. But by the time this one was over, I was stunned. Why? Because *every* one of our twenty-some students voiced the same testimony:

I've always loved the outdoors (camping/hiking/bird watching/wild ponies on Assateague/the common dolphins in the Channel Island sound/the beauty of the Ozarks). I have always felt God's presence and pleasure when I pursued those loves. But as a Christian, I didn't think I was allowed to incorporate that love or advocacy for those loves into my Christian identity. So I was really excited when you offered this course.

Every student. Every well-educated, socially active, theologically committed young adult sitting in that classroom felt they were not allowed to advocate for the beauty and sanctity of God's creation and still call themselves "Christian." And perhaps even more remarkable, the professors standing in front of those Wheaton students shared the same testimony. Why? Why has the church, historically the moral compass of our society, gotten so lost on this topic?

One reason is certainly politics. Not kingdom politics, but American and international politics. I think that most would concur that the traditional political allies of the church are not the traditional political allies of environmental concern. If you are pro-life, it is assumed that you cannot also be pro-environment. If you are a patriot, you supposedly cannot also be a conservationist. Or to be more forthright: in the United States, if you are an environmentalist, it is assumed that you are a Democrat—and Democrats, supposedly, are not pro-life. If you are a Republican, it is assumed that you cannot also be pro-environment. In other words, somehow environmental advocacy has been pigeonholed into a particular political profile and has become guilty by association. But of course, Christians are *first* the citizens of heaven, and therefore our alliances and our value system are not defined by American politics. Rather, our value system (aka "holiness") is defined by the Holy One. And as citizens of his kingdom, ultimately there is only one set of politics the Christ follower should be concerned about.

A second cause of the church's paralysis on this topic is familiar to many matters of social concern. We, the Western majority voice, are largely sheltered from the impact of environmental degradation on the global community. We don't *see* how unregulated use of land and water by big business decimates the lives of the marginalized. We have not witnessed the sterilization of the fertile fields of Punjab, India, at the hands of unrestrained industrial agriculture or the social collapse that it has caused. We have not stood on the shores of the Ganges River and seen and smelled the results of the unrelenting abuse of this immense and immensely important estuary via untreated industrial waste, raw sewage, and incomplete cremations. Our front windows do not offer us a view of the lunar landscapes left behind by mountaintop removal coal mining in Appalachia or the ragged remains of Madagascar's 88 percent deforestation—both of which have left the marginalized without recourse. As a result, we struggle to understand creation care as an expression of concern for the widow and the orphan.

Third, and perhaps most detrimental, is the theological posture taught by many in the church that the created order is bound only for destruction. Subsequently, many devoted followers of Jesus have come to believe that it is ethically appropriate to use the earth's resources as aggressively as possible to accomplish what really matters—the conversion of souls. The end result? The church, particularly the evangelical wing of the church, has inadvertently dismissed the issue of environmental stewardship as peripheral (or even alien) to the theological commitments of the Bible.

This book is my contribution to exposing and uprooting these misconceptions that have rendered the church silent on a critical concern. As a longtime professor of biblical studies, a professional exegete, an author, a theologian, and—most importantly—a committed Christian, my objective in this little book is to demonstrate via the most authoritative voice in the church's life, that of Scripture, that the stewardship of this planet is *not* alien or peripheral to the message of the gospel. Rather, our rule of faith and praxis has a great deal to say about this subject. And what the Bible has to say is that the responsible stewardship of creation is not only an expression of the character of our God; it is the role he entrusted to those made in his image.

CREATION AS GOD'S BLUEPRINT

*We all long for Eden, and we are constantly glimpsing it:
our whole nature at its best and least corrupted, its gentlest
and most human, is still soaked with the sense of exile.*

J. R. R. TOLKIEN, THE LETTERS OF TOLKIEN

I GAVE MY FIRST PUBLIC MESSAGE on the issue of environmental stewardship in 2005 at Asbury Theological Seminary's Kingdom Conference. Historically, the goal of this conference has been to engage students in larger conversations regarding Christian responsibility across the globe. Standard topics have included training for effective cross-cultural communication; messages from courageous Christian cross-cultural workers (aka "missionaries"); organizations such as Word Made Flesh and SEND International; and ministries committed to assisting orphans, refugees, and trafficked women. Never had Asbury's Kingdom Conference taken on environmentalism. But in 2005, under the courageous leadership of Professor Christine Pohl, the committee took the plunge. It was a tense moment for everyone. In central Kentucky in 2005, this was not a topic that "the church" talked about. At least not from the pulpit. But being young and idealistic, I said yes to the event and dove into the task with a full heart. I was determined to reach my audience in a fashion that would engage and challenge without offending. And in the twenty-five minutes allotted to me, I preached my heart out. To my joy, my community responded with the same—wide-open hearts. The end result? This event launched a movement at Asbury that is *still* moving forward.

We definitely had our challenges. There was more than one accusation of "hippie do-gooder-ism," there were lots of questions about finances and labor, and there was one particularly telling faculty meeting in which I had to actually *show* my colleagues where to find the numbers on the bottom of their plastic water bottles and explain what the numbers meant! But we moved forward, and we created one of the most effective institutional recycling programs I've ever seen.

The director of custodial services, Craig Reynolds, was a critical ally in this expedition into the unknown. Although he had not been socialized into institutional environmental commitments (we're talking about Wilmore, Kentucky, here), when Craig became convinced of the moral imperative, he not only joined the team but also did the hard work of designing a financially advantageous response. Craig crunched the numbers and demonstrated that recycling our copious amounts of paper was cheaper than trashing it. He found that employing a company such as Shred-it resulted in a *reduction* in labor for his custodial staff. Together we found *permanent* solutions to our particular scenario. Then came Matthew and Nancy Sleeth (of Blessed Earth fame), who further educated the community on the topic and offered their time and resources. When President Timothy Tennent arrived in 2009, he brought the seminary to a new level, making it clear that the next phase of expansion would be organized with an eye on sustainability. As a result, after "a long obedience in the same direction," this seminary has been transformed into a leading recycler in the region.

But as with so many efforts toward individual and systemic reform, the Asbury community was only able to respond to this challenge because the issue was addressed via the community's own value system. In this case, Asbury needed to hear a *biblical* argument as to why environmental stewardship matters to the kingdom.

So how does one mount a biblical argument on this topic? Like all issues of faith and praxis, to determine whether a value is biblical, it must be subjected to a survey of the biblical text. As interpreters and exegetes, we must ask the question: Do I see this particular value or precept *systematically* represented in the text as an expression of the reign and rule of God? Or is this value limited to a marginal representation in the Bible via the particularities of situational ethics? To make an argument that environmental concern is a kingdom value, the issue must rise to the level of the former—a consistent component of God's instructions to humanity, a regular attribute of God's communicated values and affections. And as all biblical theology starts in Eden, we must start our inquiry there as well.

WHAT DOES THE BIBLE SAY?

In the opening chapter of Genesis, God reveals his blueprint for creation. A close reading of this chapter demonstrates that the questions the biblical author is attempting to answer are, Who is God? What is humanity? and Where do we all fit within this cosmic plan? In figure 1, we see that the reader is offered an answer to these questions via the literary framework of a perfect "week." Here the interdependence of the cosmos is laid out within seven days of creative activity, crowned by the final day, the Sabbath. Thus, on days one through three we are offered three habitats (or kingdoms): (1) the day and night, (2) the sea and heavens, and (3) the dry land. On days four through six, the inhabitants (or rulers) of these various realms of creation are put in their proper places as well: (4) the sun and moon to rule the day and night, (5) the fish and birds to occupy the sea and sky, and (6a) the creatures who inhabit the dry land.[1]

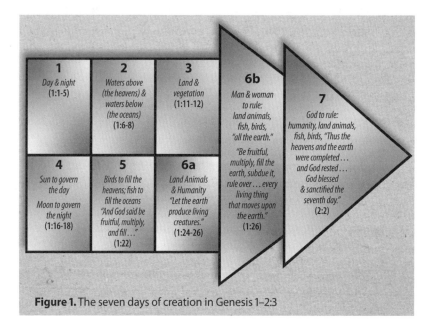

Figure 1. The seven days of creation in Genesis 1–2:3

As we consider the relationship between the first three days of Genesis's creation song, which designate the habitats/kingdoms of creation, and the final four days, which identify the inhabitants/rulers of those same realms, we find a correlation that communicates place and authority. Therefore, on day four we read that God creates the "two great lights" to "govern" (or "be lord of"; Hebrew: *māšal*) the day and night (Gen 1:14-19). On day five we read that fish and birds are created to "be fruitful, multiply, and fill" the seas and skies (Gen 1:20-23). On day six the land creatures are created to occupy the dry land (Gen 1:24-25). But as we approach the sixth day, we find that the literary structure of the piece shifts dramatically. Why? To communicate the crucial role that this stanza holds in the larger piece. Even the most casual reader can see that this day is given the longest and most detailed description up to this point. Why so much attention? Because this penultimate climax of Genesis 1 offers us the most breathtaking aspect of the Creator's work so far. On this day a *creature* is fashioned in the likeness of the Creator himself. On this day humanity (*ʾādām*) is created in the *image* of God.

> Then God said, "Let us make humanity [*ʾādām*] in our image [*ṣelem*], according to our likeness; so that they may *rule* [Hebrew: *rādâ*][2] over the fish of the sea and over the birds of the heavens and over the livestock and over all the earth and over every creeping thing that creeps on the earth." (Gen 1:26)

The profound implications of humanity (*ʾādām*) being fashioned and animated as God's physical representatives on this planet cannot be overstated.[3] Both the biblical text and its ancient Near Eastern counterparts make it clear that for humanity to be named a *ṣelem* (image) is for humanity to be identified as the animate representation of God on this planet. In essence, woman and man are the embodiment of God's sovereignty in the created order. Here male and female are appointed as God's custodians, his stewards over a staggeringly complex and magnificent universe, *because* they are his royal representatives. Like the fish and birds, humanity is commanded to "be fruitful, multiply, and fill" their habitat. But because they are the image bearers of the Almighty, they are also commanded to "take possession of" (Hebrew: *kābaš*), and "rule" (Hebrew: *rādâ*) all of the previously named habitats and inhabitants of this amazing ecosphere as well:

> God blessed them; and God said to them, "Be fruitful, multiply, and fill the earth so that you may take possession of it [*kābaš*].[4] Rule over the fish of the sea and the birds of the heavens and every living thing that moves on the earth." (Gen 1:28)

In the language of covenant, Yahweh has identified himself as the suzerain and *ʾādām* as his vassal. Moreover, Yahweh has identified Eden as the land grant he is offering to *ʾādām*.

The final stanza of the creation song introduces the ultimate climax of both the week and the message—the Sabbath day (Gen 2:1-3). This seventh day is set apart; it is sacred; it is holy. This day communicates that the universe in all of its breathtaking symmetry is finished, that the Creator is pleased, and as an expression of his good pleasure God has seated himself on his throne to revel in the beauty before him. Most important to us, the seventh day communicates that the perfect balance of this splendid and synergetic system is dependent on the sovereignty of the Creator.[5] And as God is enthroned over all the vastness of our universe on the *seventh* day, humanity's installation on the *sixth* day announces that man and woman have been appointed as the stewards of God's vast cosmos. This message is reiterated in Psalm 8, when a worshiper standing millennia beyond the dawn of creation reiterates the wonder of humanity's place in the cosmos:

> When I consider your heavens,
> the work of your fingers,
> the moon, and the stars that you have fixed in place.
> What is humanity that you should remember him?
> Or the son of ʾādām that you should care for him?
> You have made them [humanity]
> a little lower than the angels
> and crowned them with glory and splendor,
> You have made them lord [Hebrew: māšal][6] over the works of
>   your hands,
> You have placed everything under their feet
> Flocks and oxen, all of them!
> Even the wild creatures of the field!
> The birds of the heavens, and the fish of the sea,
> whatever passes through the paths of the seas!
> O Yahweh, our Lord,
> how majestic is your name in all the earth. (Ps 8:3-9)

The message in both texts is explicit. Whereas the ongoing flourishing of the created order is dependent on the sovereignty of the Creator, it is the privilege and responsibility of the Creator's stewards (that would be us) to facilitate this ideal plan by ruling in his stead. How? Like the other inhabitants of the earth, sky, and sea, the children of Adam are to "be fruitful, multiply, and fill" the earth. But as the ones made in God's image, we are also given authority over all the spheres whose creation precedes the sixth day. And like any vassal who has been offered a land grant by his suzerain,[7] humanity is commanded to "take possession" of this vast universe per the instructions of his sovereign lord (Gen 1:28).[8] In sum, humanity plays a critical role in God's blueprint for the flourishing of this majestic ecosphere in which we find ourselves. Yahweh is indeed the ultimate sovereign, but humanity has been created as his representative to serve as custodian and steward, enacting the Creator's will by living our lives as *a reflection of God's image*. We have received our authority from the Creator. We rule as he would rule. We are stewards, not kings.

Genesis 2:15 specifies humanity's task further:

> Then Yahweh Elohim took the human and put him into the garden of Eden to tend it [*lĕʿobdāh*][9] and to guard it [*lĕšomrāh*].[10]

In this second creation account, the message is repeated: the garden belongs to Yahweh, but human beings have been given the privilege to rule and the responsibility to care for this garden under the authority of their divine lord. This was the ideal plan—a world in which humanity (*ʾādām*) would succeed in building human civilization in the midst of God's kingdom by directing and harnessing the amazing resources of this planet under the wise direction of their Creator. Moreover, as those made in the image of God, humanity is literally "installed" in the garden for this very task.[11] Here there would always be enough. Progress would not necessitate pollution. Expansion would not require extinction. The privilege of the strong would not demand the deprivation of the weak.

And humanity would succeed in this calling because of the guiding wisdom of their God. As I am wont to say in my classes, God's ever-expanding universe was offered to his children such that they might always be captivated by its profound complexity, its fierce beauty, and its fragile balance. We were *designed* to love what God loves, and we were commissioned to seek the stars.

But we all know the story: humanity rejected this perfect plan and chose autonomy instead. And because of the authority of humanity's God-given position within creation, all creation paid the price for humanity's choice. Because of ʾādām, even "the creation was subjected to futility [or "frustration"]" (Rom 8:20).[12] In the words of New Testament scholar Douglas Moo, because of ʾādām's choice, the planet itself has been "unable to attain the purpose for which it was created."[13] As I discuss in my book *The Epic of Eden: A Christian Entry into the Old Testament*, the curse enacted by humanity's rebellion is not simply a list of random penalties—it is a *reversal* of God's originally intended blessings.[14] Those made in the image of God and designed to live eternally will now die like the animals. The earth, designed to serve, will now devour (Gen 3:19). The act of birth will now produce death (Gen 3:16). Adam's labor, which was intended to bring security to his family, will now be undermined by the very resources designed to provide for him (Gen 3:17-19). In other words, the perfect balance of Eden, portrayed in the seven-day structure of Genesis 1, has been flipped upside down because of the rebellion of those who were appointed to lead. The treason of God's chosen stewards has consigned all under their authority to frustration and death. This because although Adam and Eve had the authority to *make* this choice, they did not have the agency to hold the cosmos in check after making it. In an instant, God's perfect world became ʾādām's broken world—full of conflict, want, death, anxiety, and violence. And because of humanity's strategic place in God's plan, not only did this twisted existence become Adam and Eve's inheritance—it became the inheritance of all placed under their rule.[15]

WHAT WILL WE SAY?

In my experience, the body of Christ readily recognizes the disastrous effects of the fall in the arena of human relationships. Corrupt and abusive governments, bigotry and violence, the oppression of the weak and the deprivation of the voiceless—no one needs to tell the informed believer (or even most unbelievers) that these realities were not God's original intent for humanity. Nor, in my experience, does anyone need to tell the committed Christian that it is the responsibility of the church to take a proactive stand against these distortions of God's good plan. History teaches us that, at its best, the church has been among the first to identify the effects of the fall on human society and has often been the first to respond. There is a reason that most of the relief organizations, homeless shelters, hospitals, and orphanages on this planet have the words *Christian*, *salvation*, *mission*, *Baptist*, *saint*, or *cross* in their titles.[16] As Bishop Swanson of the New York City Tract Society stated in 1859 when faced with the unbearable conditions in the urban slums of an emerging America, "The Church of Christ must grope her way into the alleys and courts and purlieus of the city, and up the broken staircase, and into the bare room, and beside the loathsome sufferer. . . . For she was organized, commissioned, and equipped for the moral renovation of the world."[17]

This imprint of God's character in the heart of the true believer is why the first abolitionists were Christians; why Martin Luther King Jr. was a Baptist preacher; and why the Union Rescue Mission (currently the largest private homeless shelter in the United States) has housed itself in the bowels of LA's Skid Row since 1891.[18] We see the impact of humanity's rebellion and we know that we are called as Christians to be light, salt, and leaven in the midst of a bruised and broken world. But rarely, it seems, do we as Christians reflect on the effects of humanity's rebellion *on the garden*. And rarely, it seems, do we consider how the reality of redemption in our lives should redirect our attitude toward the same. Surely if the ultimate objective of our God is to reconcile the world to himself *through us*, this topic deserves to be on the table as well. (2 Cor 5:17-21).

> **DISCUSSION QUESTIONS**
>
> 1. What aspect of this chapter affected you the most? What was the most troubling, the most inspiring, or the most convicting?
>
> 2. In your church community, what are the main roadblocks to environmental concern and action?
>
> 3. In your own life (do your best to be transparent) what are the main roadblocks to environmental concern and action?

> # 2

THE PEOPLE OF THE OLD COVENANT AND THEIR LANDLORD

Failure to fulfill our obligations as faithful trustees of the gifts of God's creation will inevitably bring God's judgment upon us. The earth itself will rebel against our greedy and thoughtless exploitation of nature and our irresponsible fecundity.

RICHARD BAER JR., "LAND MISUSE: A THEOLOGICAL CONCERN" (1969)

RENTER OR LANDLORD: WHAT DOES THE BIBLE SAY?

Our next stop in our survey of biblical theology as it concerns environmental stewardship is the nation of Israel. Israel is critical to the discussion because it stands as the first model of God's relationship with a redeemed and landed citizenry in a fallen world. Israel understood that it was Yahweh who actually owned the land of Canaan. This emerged from their understanding that Canaan was a land grant, distributed to the tribes under Joshua; and Israel's privileges to that land grant emerged from their covenant document—their constitution and bylaws, what we know as the book of Deuteronomy.[1] The stipulations here are completely clear. If the nation will keep Yahweh's commandments, they will keep the land. Deuteronomy 4:40 summarizes the agreement with these words:

> Keep his decrees and commands, which I am giving you today, so that it may go well with you and your children after you and that you may live long in the land the LORD your God gives you for all time. (NIV)

In this book, whose legal traditions reach back into the shadows of Israel's earliest settlement, there is a continual chorus: if the people will remember the law of God and obey it, they will live and prosper; but if they forget and disobey, they will not prosper. To obey is life; to disobey, death. "So choose life in order that you may live, you and your descendants!" (Deut 30:19 NASB). The incarnation of Israel's blessing of life was the Promised Land. This is the land of Canaan that Yahweh "swore to Abraham, Isaac, and Jacob, to them and to their descendants after them" (Deut 1:8). In the language of ancient international diplomacy, the land of Canaan was, as mentioned in the first chapter, a "land grant" given by a suzerain to his vassal. And, of course, land grants could be recalled.[2] Thus, although the offspring of Abraham were invited to live on the land with joy and productivity, the book of Deuteronomy is eminently clear that, just like the Garden of Eden, the land would never be truly theirs. Rather, as the curse sections of Deuteronomy 28 unequivocally communicate, Yahweh retained the right to reclaim his land, to uproot his people "from their land in anger and fury and in great wrath, and to cast them into another land as it is this day" (Deut 29:28). Why would Yahweh, who loves his people, uproot them from their land? As Israel's story presents in graphic detail, he did so because of Israel's long-lived, oft-repeated, remorseless breach of the covenant agreement. As it was in the garden, so it was in the land of Israel. God owns the land, and it is humanity's privilege to live on it. It was God's joy to give both the garden and the land grant of Canaan to his people. It was God's intention that the land would provide for all of his people's needs. But God's people were renters, not landlords. And if they failed to remember that reality, there would be consequences.

Israel's identity as a lessee is most evident in the laws of the tithes, firstfruits, and the offering of the firstborn that populate the Mosaic covenant. Here we find that the people of Israel, much like renters, were expected to pay a percentage of their income to Yahweh via the central cult site—the tabernacle (and later the temple). In Israel's early subsistence economy, in which pastoral and agricultural goods were the mediums of exchange, this meant a percentage of their crops and flocks were to be brought to the tabernacle/temple as an offering to Yahweh. And because Israel's government was theocratic (a government actually ruled by God), an offering to Yahweh was also tribute to the king.

> Be sure to set aside a tenth of all that your fields produce each year. Eat the tithe of your grain, your new wine and your oil, and the firstborn of your herds and flocks in the presence of Yahweh your God at the place he will choose to place his name[3]—the tithe of your grain, your new wine, your oil, and the firstborn of your herd and flock—in order that you may learn to fear Yahweh your God all your life. (Deut 14:22-23)[4]

> You shall set aside each of the firstborn males that are born from your herd [cattle] and your flock [sheep and goats] for Yahweh your God. You shall not work with the firstborn of your herd, nor shear the firstborn of your flock. Rather, you and your household shall eat it every year in the presence of Yahweh your God in the place that Yahweh chooses. (Deut 15:19-20)

> This will be the priests' due from the people: when anyone sacrifices an ox or a sheep, they must give the priest the shoulder, the two

cheeks, and the stomach. You shall also give him the firstfruits of your grain, your new wine, and your oil, and the first fleece when you shear your sheep.⁵ For Yahweh your God has chosen him and his sons from all your tribes forever, to stand and serve in the name of the Yahweh. (Deut 18:3-5)

For those unfamiliar with ancient Near Eastern culture, it is best to understand Israel's tithe as a form of regular taxation. The system of offering and sacrifice served two important functions: (1) to acknowledge Israel's position as a tenant and subordinate in God's government, and (2) to address the needs of the landless among them (Deut 14:28-29; 26:12-15).⁶ But unlike the way most of us feel about rent or taxation, the law in Deuteronomy speaks of the tithe as an act of celebration as well. Here the Israelite is *worshiping* by giving thanks to God with a concrete gift that supports the staff and infrastructure of the temple/tabernacle, as well as the marginalized. As most of the meat from the animals sacrificed was returned to the worshiper, the offerings brought on this pilgrimage also provide a feast to be shared by the extended family.⁷

The law of the firstborn is quite unfamiliar to most of us. It provokes the question: What makes a firstborn special to a pastoralist? Ryan Strebeck, a past student who is currently the pastor of First United Methodist Church in Sweetwater, Texas, helped me explore this question. Ryan is a third-generation cattle rancher from eastern New Mexico and Elk City, Kansas. He has a lot of experience with firstborn calves. According to Ryan, there are no particular qualities that a firstborn calf has that following calves would not share. But what is distinct about the firstborn is the "fragile nature of one's first birthing experience." Apparently (just like humans), a cow's first calving season can be more traumatic than subsequent seasons. As a result, "Miscarrying, or 'sloughing' a calf, is more common for a heifer than [for] a five-calf cow. Mindful of this, we could probably say that a firstborn calf is more prized because of the high risk of losing that first calf."⁸ Ryan went on to explain to me that in today's market, a heifer that sloughs will likely be sent to slaughter, as she has become an economic liability. Ann Bell Stone of Elmwood Stock Farm has something similar to say about sheep. She herself is a sixth-generation family farmer from central Kentucky. Her family has kept a Suffolk/Dorset–cross sheep herd for generations. Like Ryan Strebeck, Ann Stone told me that a ewe's first birth is usually no different from its later births. *But* the first birth is a strong indicator of what sort of producer and mother the ewe will be. As sheep tend toward multiple births (an economic asset), the ewe that produces twins or triplets the first time is a ewe worth keeping.⁹ In both of these testimonials we find that the firstborn is not necessarily unique in its own particularities, but it does serve as a bellwether. So a live, healthy first birth is a great blessing to any farmer and an indicator of good things to come. Moreover, as Ann pointed out, any "firstfruit," be it produce or livestock, is a product for which the farmer has labored and waited throughout a long "hungry season." So to give the "first" away is a sign of both great sacrifice and profound confidence, sacrifice in that the farmer and his family have waited a long time for that first lamb or tomato, and confidence in that they have no real assurance, outside their trust in God, that a "second" is coming.

Returning to the laws in the book of Deuteronomy, note that the text states that the firstborn is not to be worked or shorn (Deut 15:19)—meaning that the economic benefits that might have been derived from this animal *all* belonged to Yahweh. Since the goal in a region like Palestine was for the ewes to give birth twice per year (once always in the spring), and since traditionally the best meat comes from a weaned male (two to five months old), it is probable that the sacrifice at the tabernacle/temple was of a weaned, two- to five-month-old male firstling of the flock in the fall. This particular selection would not only provide the best meat for the family feast but would cull the herd of males. The fact that the firstborn was reserved for special slaughter at the central cult site, and would need to be butchered on site for the meat to remain edible, explains the inordinately small number of bones from yearlings (sheep, goat, or cattle) recovered from Israelite villages and the large number retrieved from worship sites.[10]

What we learn here is that Israel's worship was structured around the regular acknowledgment that nothing they had was truly theirs. It all belonged to Yahweh. Here we also find Yahweh's divinely authorized taxation system—the ultimate indicator that the people of Israel were only tenants on Yahweh's land. Israel is commanded to make regular offerings of the land's produce to the divine king throughout the year *because their land belongs to God*. In fact, the old legal core of Deuteronomy is introduced and concluded by imperatives regarding Israel's tenant status. Deuteronomy 12:10-12 opens the law code with the following command to bring offerings (a form of rent and/or taxation) to the central cult site:

> When you cross the Jordan and live in the land that Yahweh your God is giving you to inherit, and he gives you rest from all your enemies around so that you may live in security, then you will bring to the place in which Yahweh your God will choose to place his name all that I am commanding you: your burnt offerings, and your sacrifices, your tithes and the contribution of your hand, and the choicest votive offerings that you vow to Yahweh. And you shall rejoice in the presence of Yahweh your God.

And Deuteronomy 26:1-11 closes the law code with a reminder of the same:

> When you have entered the land that Yahweh your God is giving you as an inheritance, and you possess it and live in it, you shall take from the first of all the produce of the ground that you shall bring in from your land that Yahweh your God is giving you, and you shall put it in a basket and go to the place where Yahweh your God chooses to place his name. And you will go to the priest who is in office at that time and say to him, "I declare this day to Yahweh your

God that I have entered the land that Yahweh swore to our fathers to give us." Then the priest shall take the basket from your hand and set it down before the altar of Yahweh your God. And you shall testify before Yahweh your God, "My father was a wandering Aramean.... He brought us to this place, and gave us this land, a land flowing with milk and honey. Therefore, I have now brought the first of the produce of the land that you have given me, O Yahweh." Then you shall set it down before Yahweh your God, and worship before Yahweh your God, and rejoice, on account of all the good things that Yahweh your God has given you and your household.

Deuteronomy, the constitution and bylaws of ancient Israel, makes it crystal clear that this good land given to God's people, as well as its produce, belongs to Yahweh. The tribes of Israel are only his tenants, who are appointed to their inherited tribal landholdings according to his good pleasure.[11]

SUSTAINABLE AGRICULTURE: WHAT DOES THE BIBLE SAY?

Now let's turn toward God's expectations of Israel as regards sustainable agriculture. In concert with Israel's understanding that it was Yahweh who actually owned the land of Canaan, a number of laws address the longevity of the land's fertility. The essential idea presented in Scripture is that each generation of Israelites is required to maintain the land in such a way that it is as fertile when they pass it on to the next generation as it was when they received it. At the core of these laws is the command regarding Sabbath rest—a mandate to humanity to regularly cease production so that the land may be allowed an opportunity to replenish itself.[12] Thus in Exodus 23:10-12 we read,

> You shall sow your land for six years and gather in its yield, but the seventh year you shall let it rest and lie fallow, so that the needy of

your people may eat; and whatever they leave the wild animal may eat. You are to do the same with your vineyard and your olive grove. Six days you shall do your work, but on the seventh you shall rest, in order that your ox and your ass may rest and the son of your female servant and the immigrant may be refreshed.

Leviticus 25:4-7 reiterates and particularizes this law:

> But during the seventh year the land shall have a Sabbath rest, a Sabbath belonging to Yahweh; you shall not sow your field nor prune your vineyard. Your harvest's aftergrowth you shall not reap, and the grapes of your untrimmed vines you shall not gather.... Rather, the Sabbath [growth] of the land shall be your food—belonging to you, your male servant, your female servant, your hired man, your temporary resident, and the immigrants among you. Even your domesticated beast and the wild animal that is in your land shall have all its crops to eat.

In other words, just like humanity, the land was to be given a Sabbath rest. In agriculture-speak, what is being described here is the practice of "fallowing"—allowing a plowed and tilled field to remain unseeded during a growing season. Not only does this ancient practice aid the recovery of soil fertility, but it also breaks the natural cycle of species-specific pests and diseases that result when a single crop is repeatedly cultivated in the same field. David Hopkins says that early Israel "undoubtedly practiced a type of short-term fallowing in which a year of cultivation was followed by a year of bare-ground fallow."[13] And as all "mixed farmers" know, a fallow field serves as excellent pasturage for livestock.[14] The remains of last year's crop provide important nutrition to this year's livestock. And in addition to aerating the soil with their foraging and hooves, these same animals generously deposit their nitrogen- and phosphorous-rich manure throughout the fields one year in advance of the next planting.[15] In all these practices, the ancient

Israelite farmer supported and enhanced the microbiology of his soil and thereby dodged the disaster of sterile land, famine, and forced relocation.[16] Keep in mind that then, as now, such farming practices limited short-term yield. In fact, Norman Yoffee theorizes that a significant contributor to the late eighteenth-century agricultural collapse in Babylonia was Hammurabi's abbreviation of fallow law in his quest to increase short-term yield.[17] In contrast, the Sabbath mandate in Israel's agricultural system limited short-term yield but helped to ensure long-term productivity.[18] Then as now, long-term soil fertility protected the poor.[19] It is very interesting to me that it was the Sabbath—this "true cessation from the rhythms of work and world; a time wholly set apart"—that established the posture of restrained production and moderate consumption that facilitated the long-term perspective commanded by God.[20]

Crop rotation was a third weapon in the arsenal of the ancient farmer who labored toward sustainable soil fertility. As any organic farmer would tell us, and as the history of urbanization in Mesopotamia dramatically illustrates, the continuous cultivation of a single crop in the same field depletes the soil of nutrients and encourages the proliferation of pests and diseases specific to that particular crop.[21] In contrast, crop rotation (particularly the rotation of certain crops, such as legumes) actually restores the soil's nitrogen content.[22] The gleaning laws (which command leaving a portion of the harvest in the field for the marginalized) also contribute to sustainability. The unharvested portion of the crop ensures something agriculturalists speak of as "crop residue," which provides essential humus to the soil.[23]

Thus we see that Israel's Sabbath law protected the long-term fecundity of the land. The sustainable farming practices this law encouraged—which limited short-term yield[24] but helped to ensure long-term productivity—were understood as "righteousness" in the Old Testament (see Job 31:38-40). Of interest is that current agricultural science is demonstrating that our modern failure to provide for long-term soil fertility is indeed leading to disaster—in the form of decreased fertility, poor nutrition, and, in many parts of the world, sterility. As we will see in a future chapter, this failure also has a devastating effect on those living on the margins.

Although I would never suggest that present-day farmers return to the agricultural methods of the Iron Age, I *would* suggest that in Israel's fallow law we find a critical ideological principle that should continue to guide our approach to the stewardship of agricultural land: It is not acceptable for any populace to take from the land everything that it *can*. Rather, as the law of Israel teaches us, God's people are commanded to operate with the long-term well-being of the land as their ultimate goal. They are instructed to leave enough so that the land might be able to replenish itself for future harvests and future generations—even though such methods will cut into short-term profits. Why? The answer offered in Leviticus is short and direct: "because I am Yahweh says your God" (Lev 25:17), and "the land is mine" (Lev 25:23).[25] In Deuteronomy, the answer comes from a different direction but is equally compelling: so that "you shall prolong your days in the land" (Deut 5:33; 30:18; 32:47). In other words: because this is Yahweh's land and Yahweh's produce, and because Yahweh intends that his land be fruitful for the next generation of tenants.[26] In sum, the constitution of ancient Israel taught that economic security or growth was not a viable excuse for the abuse of the land, and that true economic well-being would come only from careful stewardship of it.

A CASE STUDY
Industrial Agriculture and Punjab India

Compare these seemingly "primitive" biblical laws with one of our most perilous issues in the global environmental crisis—the ill-begotten gains of what William Gaud first termed the "Green Revolution" in 1968.[27]

Better known as "industrialized agriculture," this revolution was birthed in post–World War II America in response to global food shortages. The commitment was to develop and distribute high-yield cereal grains supported by synthetic fertilizers and pesticides, to modernize irrigation infrastructures, and to implement new farm-management techniques in order to increase the world's food supply. The effort was so successful that Norman Borlaug, named the "father of the Green Revolution," received the Nobel Peace Prize in 1970 for his profound contribution to ending human misery in Third World countries. But these gains have not come without a cost.

Rather, as a result of the rapid increases in the utilization of chemical fertilizers and pesticides, the implementation of hybrid cereal crops, monocultural farming (the cultivation of a single crop in a single field for extended periods of time), and systemic land overuse, the same countries that hailed Borlaug a hero are now teetering on the brink of a new agricultural disaster. A prime example is Punjab, India. In September 1997, Punjab was recognized as "one of the world's most remarkable examples of agricultural growth."[28] This region had become the largest producer of grain among all the states of India, so much so that India was actually *exporting* grain. By April 2009, however, NPR announced that "the famed 'bread basket' of India" was "heading toward collapse."[29] Thirty years after the revival, Indian farmers were using three times as much chemical fertilizer to produce the same amount of grain. And although pesticide use has steadily increased in Punjab—

Figure 2. The region of Punjab

demonstrated by numerous pesticide poisonings and rampant increase in human cancers—the insects had become resistant.[30] India was experiencing exactly what Patricia Muir, professor emeritus of the College of Agricultural Science at Oregon State University, states is the inevitable outcome of relying wholly on nonorganic, chemical fertilizers:

> Essentially, as growers add inorganic fertilizers without due attention to organics, they step onto a one-way street.... They need to add ever-increasing amounts of inorganic fertilizers to sustain their yields. It is similar to any addiction, where increasing amounts of the desired substance are required to achieve satisfaction.[31]

In addition to the impact of chemical fertilizers and pesticides, the Green Revolution in Punjab created a water crisis. The hybrid wheats and rices introduced to the region required significantly more water than the traditional species. For the locals, this means that wells that once reached groundwater at thirty feet by 2009 were being drilled to two hundred feet. By 2011, 79 percent of the groundwater assessment divisions ("blocks") in the region were identified as "overexploited" or "critical," with extraction dramatically exceeding the supply.[32] Indeed, as G. S. Kalkat, chairman of the Punjab State Farmers Commission, offered in 2009, "If farmers don't drastically revamp the system of farming, the heartland of India's agriculture could be barren in 10 to 15 years."[33] Barren. And unlike the crisis of the 1960s, this agricultural impasse will arrive after Punjab has exhausted its soil, killed off its native species, emptied its aquifers, and exponentially increased its population. What will happen on our planet when the nearly 1.4 billion people of India cannot feed themselves because their oncefertile homeland is *barren*?[34]

The drama of our current global crisis is not limited to faraway lands or international politics. Rather, as is painfully evident in the United States, the industrialization of farming has forced out small-holder farmers all over this country. At a clergy leadership conference in Mississippi in 2019 I presented on these topics. In a region of our country that traditionally would not associate the word *environmental* with holiness, I watched while pastor after pastor rose to his or her feet to grieve the social and economic collapse of their parishes as a result of industrial agriculture. Mississippi is proud of the fact that it is an agricultural state—a major producer of poultry ("broilers"), eggs, soybeans, cotton, rice, sweet potatoes, catfish, and dairy.[35] But each of these two hundred Christian leaders, who hailed from every region of the state, spoke of their family farmers as people in crisis. Forced off their land by unsustainable economic conditions, these folks had been left with no choice but to sell off their patrimony to distant corporations, and seek a new vocation. The meadows that once supported grassfed dairy cows and local produce had been replaced by acre after acre of monocrop agriculture in which fringe forest and brush areas had been removed and riverbanks exposed. Not only does this nationwide trend mark the end of a way of life; it also marks the collapse of local economies. Indeed, throughout world history, family farms have served as a safeguard against the cycle of perpetual poverty. But in our day these stewards of the soil have been left landless and jobless from the "breadbasket" of India to the delta of Mississippi. Moreover, radical increases in pesticide-related human cancers, significant nutritional loss in our food supply, and allergies connected with hybrid grains may be identified in every corner of our industrialized world.

    A final factor in this transformation of farming as we once knew it is the enormous energy consumption required for industrial agriculture. In his 2008 "An Open Letter to the Next Farmer in Chief," Michael Pollan reported that the industry of farming was responsible for 19 percent of America's annual consumption of fossil fuel. In other words, what had once been the greenest industry on the planet was quickly becoming one of the dirtiest. Whereas in 1940 each calorie of fossil fuel produced

2.3 calories of food, in 2008 the ratio was 10 calories of fossil fuel to every 1 calorie of food—and the disparity continues to grow.[36]

And so we return to the moral mandate of the Old Testament. Allow the land to rest. Don't take everything you can. Take only what you need. Leave enough so that the land might be able to restore itself for future harvests and future generations—even though such methods will cut into short-term profits. I dare to postulate that the implementation of this single, wise, "primitive" principle would have completely reconfigured the effects of the Green Revolution. Most likely this piece of ancient wisdom could have steered both the Mississippi delta and Punjab, India, away from their current agricultural emergencies.

DISCUSSION QUESTIONS

1. How do the Israelite laws of land care inform us about God's intentions for our relationship with creation?
2. If the land is actually God's, how does that affect the way you think about the land? How about your behavior toward the land?
3. Why do you think our churches, our country, and our government seem to be turning a blind eye toward the impact of industrial agriculture on the land, the farmer, and populace?
4. In a modern economy, whose job is it to protect the land? Whose job is it to protect the farmer?

3

THE DOMESTIC CREATURES ENTRUSTED TO ʾĀDĀM

What is dangerous about the consumer identity is that a consumer will rarely ask questions about the supply chain leading up to the transaction. His only concern is getting the most out of the lowest-priced product. In fact, the clients prefer to maintain their traditional role of the ignorant buyer; they want to be invisible, anonymous, and free of any culpability. Assuming a "consumer" identity is morally evasive because consumers do not feel responsible for the journey of the product. They do not ask, "Who collected the raw materials?" or "Who put the pieces together?" or "How was the product transported to the shop?" It is the responsibility of the seller to worry about all this.

Myrto Theocharous,
"Becoming a Refuge" (2016)

WHAT DOES THE BIBLE SAY?

One of the greatest gifts of the Mosaic covenant was the Sabbath ordinance. Three months free from the tyranny of Egypt, three months into their journey toward the Promised Land, Yahweh offers the children of Abraham his covenant at Mt. Sinai. The core message of this covenant? "If you will honor me as God, your only God, I will make you mine forever." The gifts of this great covenant included a new identity as a nation, the land grant of Canaan, protection from their enemies, economic security, and Yahweh's very presence among them enthroned in the tabernacle.[1] In any universe, this is an amazing offer. But pause to ponder who is standing at the foot of this mountain. This is a nation of *slaves*. These people had never known freedom. Their parents and grandparents before them had lived out their entire lives subject to their masters, laboring endlessly with no self-determination, systematically dehumanized until old age and abuse deposited their broken bodies in the ground. But in the radically different relationship offered to Israel at Sinai, a new kind of master commands his people to *rest*. Every seven days. Stop. For twenty-four hours, just *stop*. And while you are stopping, remember who you are. Stop cooking and cleaning, stop writing and networking, stop farming and building, *stop*. Why? In the words of Henri Blocher, because the Sabbath

> relativizes the works of mankind, the contents of the six working days. It protects mankind from total absorption by the task of subduing the earth, it anticipates the distortion which makes work the sum and purpose of human life, and it informs mankind that he will not fulfil his humanity in his relation to the world which he is transforming but only when he raises his eyes above, in the blessed, holy hour of communion with the Creator. . . . The essence of mankind is not work![2]

The Sabbath reminds us all what it means to be *creatures*. It reminds folks like me who work far too much that "the essence of humanity is not work."

As practicing Jews everywhere would tell us, "The Sabbath is the most precious present mankind has received from the treasure house of God."³

But we might be surprised to learn that the Sabbath is not just for humans. Rather, God says that

> the seventh day is a Sabbath belonging to Yahweh your God. You shall not do any work. Not you or your son or your daughter or your male servant or your female servant or your ox or your donkey or any of your domesticated beasts. . . . And remember that you were a slave in the land of Egypt and that Yahweh your God brought you out of there with a mighty hand and with an outstretched arm. For this reason, Yahweh your God has commanded you to keep the day of the Sabbath. (Deut 5:14-15)

The book of Deuteronomy is the constitution and bylaws of the nation of ancient Israel. In this political and theological document, according to the mandate of the covenant established on God's gift of redemption, the Israelites were to honor their God by allowing their *livestock* to rest. This should arrest our attention. The Ten Commandments tell us that humanity is *commanded* to allow the domesticated beast to rest. Why? In the words of Deuteronomy, because you were once slaves yourselves. You know all too well what it is to labor without relief, to live out your entire life captive to the whim of another, to be disallowed control over a single corner of your own existence. You know what it feels like *not* to be allowed to rest.

Just as today, in ancient Israel farm animals were maintained exclusively to provide for the well-being of humanity. The most common livestock to be found on an Israelite homestead were mixed flocks of Black Sinai goats and Awassi ("fat-tailed") sheep.⁴ Israel relied on these animals for milk, meat, cheese, goat hair (for tents, rugs, and bags), and Awassi wool for textiles of every sort. Other less obvious products of Israel's "small cattle" were fat (for candles and soap), skins (containers for wine, water, and churns), bones for tools, and parchment.⁵ The flocks were mixed partly because these two animals cohabit so well and partly to ensure the economic stability of the household. Essentially, sheep and goats were the "diversified portfolio" of the ancient pastoralist. Both animals were kept for their milk and meat, as noted above, but the Awassi sheep was by far the more valuable animal. This was in part because their meat was preferred over goat, but primarily because of their renowned fleece. Awassi wool was used for garments of every sort, and we have extensive records of the lucrative exchange of this textile dating back into the third millennium BCE.[6] If life went smoothly, Awassi sheep brought their owners significant economic returns—the "stocks" of an ancient investment portfolio.

The Black Sinai goat was not as valuable. A reliable provider of milk and meat, yes, but its coarse hair was utilized only for tent curtains, bags, and other "rough" textiles. So here's the catch. Although the Awassi sheep were the more lucrative investment, they were also the more vulnerable asset. Awassi are picky eaters, sensitive to drought and heat, and pretty much defenseless against predators. As Timothy Laniak details, they have no biting teeth or claws, they get lost easily, and they are terribly nearsighted.[7] Awassi also panic easily. So when one of these sheep gets lost (as they are wont to do), they typically hunker down and begin to cry—a very effective way to locate the nearest predator. Their only self-protective instinct is to huddle (which of course reminds me a lot of committee meetings). Also not an ideal strategy (the huddling, not the committee meetings). Black Sinai goats, on the other hand, are tough as nails. These goats have been indigenous to the region for centuries. They are quite independent and are very capable of returning to an undomesticated state if need demands. The Sinai goat has an extremely high tolerance for heat and drought—these animals will eat just about anything and can consume as much as 35 percent of their body weight in water in a matter of minutes. Even during the hottest part of the season, they only need to be watered once every four days.[8] Goats were therefore the "bonds" of the Israelite farmer's portfolio. Even if the market went south, the goats didn't.

Another reason Israel's shepherds kept goats among their mixed flocks is that goats are smart—exasperating, but smart. Whereas a sheep, when confronted by a predator, will stand there and die a slow and horrible death, a goat will fight back, run, or even climb a tree if the opportunity presents itself. A herd of sheep will follow wherever they are led (including over the edge of a steep ravine); a goat will come up with a better plan. How these realities affect our interpretation of Jesus' parable about separating the sheep and the goats on the final day of judgment (Mt 25:32-33) I will leave to your imagination, but from an economic perspective, these 40/60 mixed flocks of caprids (the technical word for sheep and goats) could be found in any Israelite household and greatly outnumbered any other livestock in Israel's world.

Bovines (cattle and oxen) were another essential member of the Israelite family's workforce. These animals were far too expensive for the common person to eat and were instead utilized for the cultivation of grain. In the small-holder farms of the central hill country, the cereal crop was fundamental to the survival of man and beast.[9] As a result, the Iron Age farmer relied heavily on the labor of his beast for the long and arduous task of plowing his fields, moving the harvest, and extracting the precious grain from the stalks in which it grew (i.e., threshing; see fig. 3).

Once cleaned and stored, wheat and barley served as the primary food source for the community. And in Israel's subsistence economy, every kilo counted. A subsistence economy is defined most simply as an economy where everyone is barely making it.[10] Surplus is the anomaly. Israeli archaeologist Baruch Rosen has done some fascinating work in order to determine exactly what "just making it" looked like in Israel during the settlement and early monarchic periods (the Iron Age I, 1200–1000 BCE). His data derives from the archaeological remains of the extant Iron I sites in the hill country. Working from the material remains of these communities, he calculated to juxtapose the population estimates with harvest predictions, and his conclusions quantify exactly how many

Figure 3. A team of oxen threshing wheat

calories it would have taken to sustain a typical village of one hundred souls in a normal agricultural year.[11] To our surprise, Rosen's research indicates that the average Israelite village experienced a shortfall of fifteen million calories *per year*.[12] Anticipating that the average family included five people, this shortfall would amount to sixty days of food per family per year. Although this sort of "hungry season" is not a surprise to the anthropologist (many agricultural communities experience the same), it certainly helps the modern reader to humanize the experience of our biblical ancestors. Sixty days a year short on your family's *essential* food supply. Rosen hypothesizes that most families mediated this shortfall by truncating daily rations, attempting to raise and store more grain, and/or by slaughtering additional animals from the flock. But of course more grain required more land and seed, and slaughtering an extra animal would put the farmer behind the eight ball *next* season. Hunting was an option as well, and Deuteronomy often speaks of the wild gazelle as a part of the Israelite diet (think Pennsylvania dairy farmers and white-tailed deer; Deut 12:15, 22; 14:5; 15:22). But whatever potential solutions a farmer contrived, the point here is that when harvest time finally came, our heroes were counting and conserving every kernel of wheat and barley coming in from the fields. Knowing these economic realities, let's pause over Deuteronomy 25:4: "Do not muzzle an ox while he is threshing [the grain]." I am guessing that this simple agricultural law is one very few people lose sleep over. But with the information above, I hope you are beginning to realize that when God commands the Israelite *not* to muzzle his eight-hundred-pound working bovine, he is talking to a man who is hungry. And the five to seven pounds of grain that an ox could consume over a single day of threshing made a difference.[13] And what about the fact that threshing could go on for days? Yet God commands his farmers to allow the beasts who served them the opportunity to enjoy their life and work, to benefit from the fruit of their labors, to celebrate the harvest—even when the farmer knew such a privilege for his beast would cut into his family's essential food supply.

CASE STUDY
Mass-Confinement Animal Husbandry, aka "Factory Farming"

How would these laws of Sabbath rest and threshing reflect on the treatment of livestock in the United States? I speak of the billions of animals who are currently serving us on America's factory farms. Factory farming is the practice of raising livestock in confinement at high stocking density, where the farm operates essentially as a factory whose end product is protein units. Confined animals burn fewer calories, their excrement is mass-managed (or mismanaged, as many would argue),[14] and their fertility and gestation is fully controlled.

America's most lucrative agricultural product is pigs. Confinement for these animate (and highly intelligent) creatures has been distilled into an exact science: twenty 230-pound animals per 7.5-foot-square pen.[15] The metal-barred pens in which America's 74.6 million pigs[16] are presently housed may be found all over our country systematically mapped out within enormous metal frame structures—the most popular being the 40-, 60-, and now 122-foot-wide "wean-to-finish" buildings that confine these creatures from birth to slaughter.[17] Here pigs live out their entire lives housed on concrete and metal-grated flooring in climate-controlled conditions, never actually exposed to the light of day.[18] These animals are sustained in such crowded and filthy conditions that movement is difficult, natural behaviors are impossible, and antibiotics are essential to the control of infection. Sows, typically weighing in at 500 pounds, are housed separately—in 7-foot-by-22-inch metal gestation crates. These animate creatures *stand* in the same position, shoulder to shoulder, unable to turn or lie down for the entirety of their 112- to 115-day pregnancies. The obsessive behaviors and injuries sustained from this confinement are heartbreaking. As reported by numerous animal-welfare groups, the confinement and concrete flooring produce chronic pain, foot injury, joint damage, and massive muscle and bone-density loss. When the sow does attempt to lie down, skin lesions and leg injury are inevitable. Urinary tract and respiratory infections result from constant exposure to their own feces and the sows' lack of water intake due to what most diagnose as chronic depression/frustration.[19] Sows are artificially inseminated to deliver an average of eight litters (2.1 to 2.5 litters per year), litters inflated by means of fertility drugs to exceed the sows' natural carrying capacity. As of 2019, there were 6.41 million head of breeding sows in the United States—the vast majority of which are confined in gestation crates.[20] One week before delivering, sows are transferred to "farrowing crates"— slightly larger barred enclosures that allow the sow (at last) to lie down, with an additional eighteen inches for the piglets to nurse prior to their forced weaning at twenty-one days, when the cycle begins all over again. And when the animals cannot survive these conditions? A staple of the confined hog diet is the rendered remains of their deceased pen mates.[21] Surely, if God is offended by boiling a kid in its mother's milk (Deut 14:21), we should be concerned that dead sows (and piglets) are routinely ground up and fed to their offspring.[22]

And what of poultry? There are two categories of factory-farmed poultry in our country. The first are raised for egg production; the second are "broilers," which are bred for meat. According to the USDA's Animal and Plant Health Inspection Service, "hatchery statistics for 2010 list 9.28 billion broiler-type chickens hatched, 489 million egg-type chicks hatched, and 281 million poults hatched in turkey hatcheries."[23] Most of the 489 million hens hatched to provide our eggs will live out their entire lives confined in ten-inch "battery" cages, stacked one atop the next, row on row, in windowless warehouses. The Humane Society reports that on average, each laying hen is afforded 67 square inches of cage space—less space than a sheet of paper.[24] Among the most intensively confined animals in agribusiness, these hens will never nest, perch, dustbathe, peck the ground, walk, or even spread their wings. Nobel Prize–winning Austrian zoologist, ethologist, and ornithologist Dr. Konrad Lorenz states,

> The worst torture to which a battery hen is exposed is the inability to retire somewhere for the laying act. For the person who knows something about animals it is truly heart-rending to watch how a chicken tries again and again to crawl beneath her fellow cagemates to search there in vain for cover.[25]

As they are immersed in their own feces (and that of their caged comrades), not only is this practice inhumane, but the implications for human health are frightening. In 1999 the European Union banned battery cages, allowing their farmers a twelve-year phase-out period, which is now complete. But the United States has not followed suit.[26] Moreover, as federal animal-protection laws do not apply to chickens on-farm, and the government does not monitor animal welfare on-farm, these animals have no advocate.[27]

Figure 4. Egg-laying chickens in battery cages

"Broilers" are birds produced for meat—breast meat in particular. Chickens raised for meat today grow at a staggering rate—300 percent faster than those raised in 1960. According to the National Chicken Council, in 1925 a chick progressed from birth to slaughter in 112 days. In 2011 a chicken went from birth to slaughter in 47 days. And whereas slaughter weight for the average chicken in 1925 was 2.5 pounds, in 2011 it was 5.8 pounds and continues to rise.[28]

How has the industry achieved such amazing (and profitable) results? One reason is the structure of the industry. Forty companies own almost all the farmed chickens in our country. As the ASPCA reports, these same companies also own the hatcheries, feed mills, slaughterhouses, and processing plants.[29] This monopoly allows corporations such as Tyson and Pilgrim's Pride to control every aspect of production from hatching to slaughter.[30] These poultry-processing companies provide the genetically altered birds, customized food, and select equipment to the "growers" who do the actual rearing of the animals. The growers are required to follow precise instructions and are paid on a performance-based incentive system. In other words, the farmers who most efficiently convert the issued feed into weight gain (and therefore the highest weight of birds delivered to the processing plant) are paid the most. The exposés of the inhumane conduct that this system breeds are deeply disturbing for anyone who cares about animals . . . or humans.[31]

As a result, unlike the chickens our grandparents ate, today's fast-growing birds (in particular the "Cornish Cross" breed) have been genetically designed so that their breasts grow faster than the rest of their bodies. In addition to altered genetics, their natural habits are manipulated by continuous lighting such that they feed continuously.[32] Thus, by "harvest" the birds have become so top-heavy that they can no longer stand or walk. Their legs and organs cannot support their enormous, distorted bodies. As the industry norm is to pack as many chickens as possible into each pen, many of these animate creatures find themselves literally trapped in their own bodies, stranded on the floors of their own pens, unable to reach water or food.

After only a few weeks, there is evidence that the birds' skeletons and organs cannot keep up: their hearts, lungs and legs strain to work under severe pressure, causing severely low stamina, shortness of breath, trouble standing and walking, collapse and even congestive heart failure.... Overweight, weak and with almost no room to move, birds spend up to 90% of their lives lying down in their litter, a combination of bedding and excrement.... It is not surprising that as birds lie with open wounds directly in their own waste, in which live bacteria is known to survive, their sores can become "a gateway for bacteria which can cause...secondary infections (staphylococci spp. and e. coli)"—some of the most notoriously common foodborne pathogens that are often traced back to chicken farms.[33]

Not only should these realities shame us; they should frighten us as well. This genetically altered, factory-raised bird, which has spent the bulk of its existence lying in its own feces, is the standard source of all the chicken we buy, eat, and feed our children.

What is the rationale for this new version of "farming"? May I say up front that I've yet to meet a real farmer who would choose to abuse his animals. But an economy that leaves no space for humane animal husbandry can create strange bedfellows. All of these innovations make these production units (i.e., animate creatures) easier to manage, maintain, medicate, and slaughter. Thus, profit and the rapidly escalating market for meat for human consumption, in the Third World in particular, is named as the rationale for mass-confinement animal husbandry.[34] As Matthew Scully painfully illustrates in his 2002 exposé of the industry, *Dominion: The Power of Man, the Suffering of Animals, and the Call to Mercy*, we have seen a revolution in our country in the last several decades regarding the production and consumption of meat. We eat more meat, more cheaply, than any other generation in history.[35] As a result, in the United States, the abuses to which domesticated animals are routinely subjected are nearly too horrific to report. I find it difficult to believe that this is what Yahweh intended for the creatures he entrusted to ʾādām.

Slaughter. Consider as well the complex legal structures that accompany the slaughtering of animals in ancient Israel. God's people were certainly allowed to slaughter and eat the animals they raised, but any domestic animal had to be taken before the priest first. According to Leviticus 17, this practice ensured that the animal's *nepeš* (its life) had been considered.[36] In Israel, the life of a domestic beast was not to be taken without thought or without mercy. Deuteronomic law required that even the wild gazelle be slaughtered with due care (Deut 12:15, 22; 14:5; 15:22).[37] The Talmud (the central collection of Jewish civil and ceremonial law) mandated that the method of slaughter for the Jews of the second century and beyond be as humane as possible. In his commentary on the book of Leviticus, Jacob Milgrom states, "All of these [details] clearly demonstrate the perfection of a slaughtering technique whose purpose is to render the animal immediately unconscious with a minimum of suffering."[38] As regards the slaughterer himself, the Talmud requires that "by virtue of his training and piety, his soul shall never be torpefied by his incessant butchery but kept ever sensitive to the magnitude of the divine concession in allowing him to bring death to living things."[39]

Reflect on these Israelite laws in comparison with the assembly-line approach we employ in the raising, slaughtering, and mass-marketing of animal flesh in America. Few of us realize that animals used in agriculture have almost no legal protection. More than 95 percent of them—poultry—aren't even included in the regulations implementing the federal Humane Methods of Slaughter Act, the law that requires an animal to be rendered insensible to pain before it is killed.[40] As regards the cattle industry, Scully reports that whereas in 1990 the typical American slaughter plant operated at fifty kills per hour, by 2002 plants were running at three to four hundred per hour. How does one go about slaughtering four hundred eight-hundred-pound bovines in an hour? As Martin Fuentes, an Iowa Beef Packers worker, told the *Washington Post*, "The line is never stopped simply because an animal is alive."[41] Ramon Moren—whose job is to cut off the hooves of strung-up cattle passing by at 309 an hour—reports that although the cattle are supposed to be dead when they reach him, often they are not: "They blink. They make noises. The head moves, the eyes are open and still looking around. They die piece by piece."[42] In contrast, at every juncture, Israel was constrained to consider the life of the animal that served them and which they consumed by covenant law, even though such considerations were costly in time and resources.[43]

Good news. The good news is that there are organizations all over the United States working hard to get information and images regarding our current practices to the American public. Why? Because these folks believe that once we *see* where our eggs come from, and the gross exploitation of the animals that are providing the meat on our dinner tables, we will put pressure on the industry to change—pressure from our refusal to participate with our finances and pressure through legislation. The American Society for the Protection of Cruelty to Animals, long focused on the well-being of pets, has returned its focus to the protection of the working animal. As a result, its website is now populated with white papers and reports exposing the inhumane practices of these industries and offering the average citizen a means by which to help bring about change.[44] Check it out.

Bad news. The bad news is the enormous corporate influence behind these industries and what have come to be known as "ag-gag" laws. Ag-gag laws are a growing body of state-level legislation designed to criminalize whistleblowers who expose inhumane practices in animal agricultural facilities by photographing or videoing what is going on behind closed doors.[45] At this point, more than half of the state legislatures in our country have attempted to pass some version of these laws, and the practice is spreading to other countries as well.[46] As a result, in seven states it is currently illegal to take a photo or video of a slaughter plant. Why? Because images of this sort put us face-to-face with practices that are intolerable.

One particularly arresting story involving ag-gag laws is of twenty-five-year-old Amy Meyer, who attempted to take a cellphone video outside the Dale T. Smith and Sons Meat Packing Company in Draper, Utah.[47] A Draper resident, she had driven past the meat packing facility many times, but on February 8, 2013, she decided to stop. Standing on a strip of public land outside the barbed-wire fence enclosing the facility, she began filming. Her brief video shows the open doors of the facility facing the public highway and the cows being led onto the plant's assembly line. She recorded a forklift pushing a live, downed cow outside the building. Within seconds of Meyer's beginning to film the slaughterhouse and narrate the events, the facility's operator pulled up in a truck and informed her she was breaking the law. When she pointed out that she was on public land, he called the police. A few minutes later, *seven* squad cars arrived. "The officers would all go to Brett Smith [the facility's operator] first and shake his hand," Meyer recounts.[48] Meyer was questioned, accused of trespassing (even though her film made it clear she had not), and threatened with charges.[49] Eleven days later, Meyer learned that she had been charged with "agricultural operation interference," a class B misdemeanor that carries a maximum six-month jail term.[50] Lucky for Meyer, the story was picked up by the local and national media and the charges were dropped. All this for filming what is visible to any passer-by with a cellphone? I'm grateful to say that as ag-gag laws have become more public, so has the opposition. It is my hope the same will happen when the practices of mass-confinement animal husbandry make it into the public eye. My question is, Where will the church be when this happens?

CASE STUDY
Jim Goodman

One last testimony regarding the impact of factory farming is that of Jim Goodman—a dairyman with forty years in the business. As I have already noted, a primary casualty of factory farming is not the animals; it is the farmer. We have reached a place in our society where the small family farm simply is not economically viable. The chicken farmer who values his charges as animate creatures cannot compete against Tyson. The dairy farmer who knows his cows by name cannot match resources with the corporate dairies that milk hundreds, even thousands of cows on mechanical carousels. (Although, according to the children's website Dairy Discovery Zone, we should all be reassured that the "wholesome, nutrient-rich milk" retrieved via these mechanical carousels "is never touched by human hands."[51])

Jim Goodman launches his 2018 testimonial in the *Washington Post* by saying, "I sold my herd of cows this summer. The herd had been in my family since 1904; I know all 45 cows by name. I couldn't find anyone who wanted to take over our farm—who would? Dairy farming is little more than hard work and possible economic suicide."[52] Goodman speaks of how he couldn't watch while his cows were loaded on the truck. So he milked them for the last time, left the barn, and let the truckers take them away. "Being able to remember them as I last saw them, in my barn, chewing their cuds and waiting for pasture, is all I have left."[53] Goodman is clearly a capable businessman. He understands the realities of supply and demand, farming legislation, and politics. But he also understands that farmers are becoming obsolete. "When I started farming in 1979, the milk from 45 cows could pay the bills, cover new machinery and buildings, and allow us to live a decent life and start a family."[54] Being a good businessman, Goodman survived the 1980s—a decade in which 250,000 farms in the upper Midwest went under, and more than 900 farmers committed suicide. But Goodman says the current crisis is bigger than he is. And so along with the 665 other dairy farms in Wisconsin that closed between 2017 and 2018, Goodman's farm, his home, and his family's legacy are "done." Goodman credits this ending to "ineffective government subsidies and insurance programs," which "are worthless in the face of plummeting prices and oversupply.... The despair is palpable; suicide is a fact of life." Of course, when a community of farms goes down, so do the local businesses that built their lives around those farms—the cafés, grocery stores, schools, and churches associated with the farmers and their families. Essentially, when a constellation of farmers are forced to call it quits, the infrastructure of a community collapses.

> With fewer farms, there are fewer foreclosures than in the 1980's. But watching your neighbor's farm and possessions being auctioned off is no more pleasant today than it was 30 years ago. Seeing a farm family look on as their life's work is sold off piece by piece; the cattle run through a corral, parading for the highest bid; tools, household goods and toys piled as "boxes of junk" and sold for a few dollars while the kids hide in the haymow crying—auctions are still too painful for me.[55]

When I hear Jim Goodman's testimony, and so many like it, I hear the voices of the widows and the orphans of the modern era. And I wonder, who will speak up for them?

According to the US Department of Agriculture, the number of dairy farms in the United States dropped from nearly 650,000 in 1970 to 40,219 by the end of 2017. That is a staggering differential. How can we explain this? Cows are actually producing more milk, and Americans are consuming more milk than ever before. The answer is that the number of cows on American dairy farms has skyrocketed. Whereas in 1987 half of American dairy farms had eighty or fewer cows, as recently as 2012, that figure had risen to *nine hundred* cows. Indeed, the dairy cows of this generation are now consolidated (warehoused) on bigger, more "efficient" farms. But one farmer does not build a relationship with nine hundred cows—a corporation does.

CONCLUSIONS

I began this chapter with a quotation from Myrto Theocharous, a brilliant young Old Testament scholar teaching in Athens, Greece. The quotation emerged from a plenary presentation she offered at the national gathering of the Evangelical Theological Society in the fall of 2015. Here Theocharous defined *consumerism* and explored how a consumer mentality intersects with a holy life. As in her published article, she states that the only true concern of a consumer mentality is getting the lowest price for the best product. This doesn't necessarily sound bad to me—in fact, this is how I typically shop. It is probably how you shop too. But in the plenary gathering of the Evangelical Theological Society, when Theocharous moved the conversation regarding consumerism into her own ministry setting, the landscape for all of us in the audience changed abruptly. You see, Theocharous doesn't exactly work in retail. She works with the young women who labor in the brothels of Athens. Consider the quotation (in its larger context) again:

> What is dangerous about the consumer identity is that a consumer will rarely ask questions about the supply chain leading up to the transaction. His only concern is getting the most out of the lowest-priced product. In fact, the clients prefer to maintain their traditional role of the ignorant buyer; they want to be invisible, anonymous, and free of any culpability.[56]

I can testify that on that November afternoon, in the grand ballroom of the Hilton Hotel in Atlanta, Georgia, we all got a worldview makeover. Because when we think in terms of getting the lowest price for the best product in Theocharous's context—there is no question in our minds that such behavior is immoral. How can we *not* be concerned with how the prostitute or actress in a pornography film came to be in the industry?

According to Theocharous, assuming a "consumer" identity is morally evasive because consumers do not feel responsible for the journey of the product. They do not ask, "Who collected the raw materials?" or "Who put the pieces together?" or "How was the product transported to the shop?" It is the responsibility of the seller to worry about all this.[57] In the sex industry, I doubt anyone would challenge Theocharous's thesis—of course we are responsible. Our capitalist economy and consumer culture cannot absolve us of such license. So now for the hard question:

How about the "consumer culture" that facilitates our purchase of milk, meat, and eggs? Are we morally responsible for how these "products" come to us—who and how the raw materials were collected? Do we have an obligation to the creatures who produce them? Or are we free to claim absolution as to the "journey of the product" in our quest to get the best product at the lowest price? In that Kingdom Conference sermon I gave oh so many years ago at Asbury Theological Seminary, I risked the question: "Have you ever considered the *life* of the styrofoam and cellophane packaged chicken parts you purchase at Walmart every week?" Israel was constrained to do so by covenant law.

The laws of the Old Testament make it clear that the people of Israel were not free to be simple "consumers" of the domestic creatures entrusted to them. They were commanded to honor their God by honoring their beast—a Sabbath's rest, humane treatment in its life and work, a share of the harvest, slaughter with dignity and compassion. As the people of God today, can we offer our God anything less as regards the creatures entrusted to us?

DISCUSSION QUESTIONS

1. Why might the ancient law of Deuteronomy be so focused on animal welfare?
2. Why is it that we in the twenty-first century seem to show so much less interest in the welfare of our livestock?
3. Do you come from a farming family? Perhaps a 4-H family? How do you think farmers who have chosen to stay in the business feel about the practices described above?
4. Why do you think local towns embrace mass-confinement animal-husbandry companies?
5. Do you have any new thoughts on the "consumer mentality"?
6. Why do you think our legislators have allowed the collapse of the family farm and all the village economies that have collapsed with it?

4

THE WILD CREATURES ENTRUSTED TO ʾĀDĀM

> *We in the industrialized world have allowed our appetites to outrun both our resources and our humanity.... Our sages did not condemn materialism.... But they were acutely aware, at the same time, of the need for balance, a balance we scarcely any longer recognize.*
>
> Daniel Swartz, "Jews, Jewish Texts, and Nature: A Brief History" (1994)

In chapters 38 and 39 of the book that bears his name, Job, the long-suffering servant of the Most High, is hammered with a series of questions from on high. The intent of the interrogation? To remind Job that he is creature, not Creator.

Have you ever in your life commanded the morning, or caused the dawn to know its place?... Have you entered into the springs of the sea, or have you walked in the recesses of the deep?... Can you stalk prey for a lioness, and satisfy the young lions' appetites as they lie in their dens or crouch in the thicket?... Do you know when the mountain goats give birth? Have you watched the calving of the does? Have you counted the months they carry their young? Are you aware of the time of their delivery? . . . Is it by your understanding that the hawk soars, stretching his wings toward the south? Is it at your command that the eagle mounts up, and makes his nest on high? (Job 38:12, 16, 39-40; 39:1-2, 26-27)

When I hear these questions voiced, I echo Job's response: surely not I. I can hardly understand these marvelous things, let alone mimic or duplicate them. Only the master of the universe can call an eagle from its perch or command the dawn to take its place. So I, like Job, respond to creation with praise for the Creator. When I stand at the ocean's edge and feel the spray of its raging force on my face, when the wind silences me, when I am privileged to hold a wild creature in my hands, my heart cries out with the psalmist, "O Yahweh, our Lord, how majestic is your name in all the earth!" (Ps 8:9).

Why is my heart moved to worship by the splendor of an eagle on the wind, the staggering realities of life in all its complex forms? Why do I sit in front of my television watching *March of the Penguins* with my seven-year-old and find myself in awe of a God who could instill in the heart of a *penguin* a level of self-sacrificial obedience that puts this believer to shame? The answer is most simply because the cosmos, in all its beauty and complexity, is a reflection of the God who made it. And I am made in the image of that same God. So the part of me that remembers Eden is stunned into silence when I step into a meadow and unexpectedly lock eyes with those of a wild creature. The part of me that can still hear the echo of the Maker's song laughs with joy when I see dozens of common dolphins race across the channel so they can "bow ride" off the front of my whale-watching boat—diving in perfect synchronization, breathing as one, sleek and shimmering in the sun. There is no question that our God has graced this planet with creatures of such amazing beauty and skill that it nearly defies imagination. Wild and fierce and yet oh so vulnerable, these are the creatures that God has entrusted to ʾādām.

WHAT DOES THE BIBLE SAY?

So what do the Scriptures have to say regarding the wild creatures that inhabit this planet with us? The Bible makes it very clear that even in a fallen world, God rejoices in the beauty and balance of his creation. It is Yahweh who "sent out the wild donkey free" and "gave to him the wilderness for a home" (Job 39:5-6). In the flood narrative, although God judges the world because of its corruption, he rescues animal-kind along with humankind, and his re-creational covenant is with "every living thing . . . the birds, the domestic animals, and every wild creature of the earth" (Gen 9:10-11). In the elevated language of Psalm 104 we read,

> He is the one who sends forth the springs into the wadis;
> between the mountains they flow;
> giving drink to each of his wild creatures. (Ps 104:10-11)

The Scriptures teach us that Yahweh has designed our ecosystem *so that* his wild creatures will have the food, water, and habitat they need to survive and flourish.

> The trees of the LORD drink their fill—
> the cedars of Lebanon that he planted.
> Where the birds make their nests,
> and the stork builds its house in the cypresses.
> The high mountains are for the mountain goats;
> the crags are the refuge of the rock badgers. (Ps 104:16-18)

Clearly, God has intentionally provided this vast array of creatures the habitats necessary to their survival. But as any environmentalist would tell us, the single greatest cause of the extinction of an animal species is the destruction of its habitat. And in America we are presently devouring nearly two million acres a *year* in the noble quest for urban sprawl.[1] As a result we are also experiencing a species extinction rate of as much as one thousand times the historical loss ratio.[2] It would seem that "we in the industrialized world have [indeed] allowed our appetites to outrun both our resources and our humanity" and have reached a point where we need balance, "a balance we scarcely any longer recognize."[3] And the fact that the wild animals' habitat was designed and given to them by our God should inspire us to reconsider our reckless behavior.

The territory we know as "Israel" is in reality a very small region. But it is graced with dramatically distinctive geographic and climatic variety. Here we find both temperate and tropical zones, the Mediterranean to the west and Red Sea to the south, deserts east and south, the Central and Transjordanian mountain ranges, the Jordan Rift Valley, the Dead Sea (the lowest land elevation on earth), and the great east-west passageway known as the Jezreel Valley. This region also serves as a land bridge between two continents—Africa and Asia. In other words, the "Promised Land" is not just a destination; it is also a very important migratory route. As a result of its diverse ecosystem and critical role in animal migration, the variety of species that have inhabited this small territory is extraordinary. The white oryx, the Syrian brown bear, the Asiatic lion and cheetah, and the Syrian wild ass were here before their habitats were destroyed.[4] The jungle cat once occupied the now-defunct Hulah River basin. The sand cat, wild cat, caracal, and leopard were once found here as well. Aurochs, bubal hartebeest, the Nile crocodile, and the hippopotamus once inhabited the Jordan River and its densely vegetated alluvial plain. Some believe the Syrian elephant passed through this region. The acacia and dorcas gazelle, Persian fallow, roe, and red deer, the Nubian ibex, and wild boar were all food sources (see Deut 12:15, 22; 14:5; 15:22).[5] The Arabian and north African ostrich, hoopoe bird, eagles, hawks, owls, and songbirds too numerous to name either currently inhabit this region or are known via zooarchaeology (bones retrieved from excavation). There are more snakes in the southern Levant than you want to know about.[6] Lions, leopards, bears, jackals, foxes, and wolves have all been identified either through biblical references or excavated remains.[7] Then there is the Eilat Coral Beach Nature Reserve just off the southern tip of Israel, filled with a staggering array of sea life. This is the sort of space that the Nature Conservancy would purchase if it could.

Based on the larger history of the area, we can safely assume that in the early stages of Israel's settlement and urbanization (1200–1000 BCE, the Iron Age I) the Israelites did not yet constitute a serious threat to this complex Levantine ecosystem. Rather, early Israel colonized the highlands in small villages of two hundred to three hundred people, organized around extended families of fifteen to twenty persons, in a closed and reciprocal economy. Each of these small villages was absorbed with the all-consuming task of survival.[8] Dry farming of grain, grapes, and olives on terraced hillsides, complemented by ongoing mixed animal husbandry of sheep and goats, characterized this subsistence economy.[9] These simple settlements display minimal permanent architecture. Domestic dwellings encircled the village to shield the people and their flocks from danger; storage was limited to small, lined family silos for grain and plaster-lined cisterns for water; there was no monumental architecture.[10] This was a "closed" economy that had limited contact with the outside world. Exchange was chiefly barter-based and in-kind, and metal of all sorts was scarce.[11]

Thus we can assume that when the societal regulations of Deuteronomy were conceived, the habitat of the wild creatures of the Promised Land was not yet under undue stress from human settlement. But Deuteronomy 22:6-7 offers us a very curious little law making it clear that even in this stage of human development, the preservation of the indigenous species of the region was a priority:

> If you happen on a bird's nest in front of you in the road, or in a tree, or on the ground, with young ones or eggs, and the mother sitting on the young or on the eggs, do not take the mother (who is sitting) on the young. Rather, you will surely shoo the mother away, and the young you may take for yourself, in order that it may be well with you and that you may prolong your days.

Many have identified this law as a *pars pro toto*: one expression of a larger principle offered as a representative of the whole. Several have also identified it as *analogia*: a vehicle of Wisdom literature that formulates an abstract idea by means of a practical example.[12] And most have identified in this passage an analogy to Deuteronomy 20:19-20—the law sparing fruit trees during siege warfare, which I'll address in the next chapter. The common idea between these texts is the preservation of the means of life.[13] In other words, the idea of sustainability. In this law, the practice of taking both mother and offspring is censured in that such a practice will eventually lead to the extermination of a particular species in a particular place. Moreover, as famed Jewish scholar Jeffrey Tigay points out, the phrase "mother with her children" often appears in descriptions of warfare as a byword for wanton killing.[14] So we see that even at the earliest stages of its urbanization, Israel is commanded to live in a sustainable fashion in its engagement with the wild creature.

Also interesting to us is that the group in Israel's world best known for wanton killing was the Neo-Assyrian Empire. The Neo-Assyrians were the first true world empire. Launching under the very capable leadership of Tiglath-pileser III in 745 BCE, this completely militarized state assembled the largest and best-equipped fighting force the world had ever seen. They were notorious for their brutality, economic oppression, insatiable appetite for power, and ambition to rule the known world.[15]

Most nations in the path of Neo-Assyria's expansion chose to cooperate with the crown and were transformed into vassals of the empire. This meant that the local ruler retained his office and national boundaries, but was charged with an annual tribute of significant proportions that often placed the economic stability of his state in jeopardy. A vassal king was expected to open his borders to the unimpeded passage of the Neo-Assyrian army, feed and house its forces on their way, and supply conscript soldiers as well. Ultimately, all roads led to Nineveh, and all resources flowed there as well.[16] Less cooperative local leaders were stripped of their thrones, either exiled or executed, and replaced with a man of the emperor's choosing. If rebellion continued, the local government was obliterated, an Assyrian governor appointed, and the territory absorbed into the empire as a province.[17] This final stage often saw the exile of the indigenous populace and the gifting of the conquered territory to another people group. This strategy stripped the conquered nation of its will to rebel by relocating the bulk of its population elsewhere and repopulating the homeland with a foreign people group. In this fashion, "national identity was lost, dissident factions dissolved, and the new heterogeneous populace in both the old and new territories were left with survival as their only objective and Assyria as their only lord."[18]

One of the more undesirable qualities of the Assyrian royalty was that they traditionally advertised their "right to rule" via images of an array of violent acts. One of these was the kingly act of slaying a wild (Persian) lion. The "press release" version of this royal accomplishment is depicted in figure 5. Here the Neo-Assyrian monarch is displayed in all his heroic splendor, killing a male lion alone and unaided. What the public didn't know is how this image came to be, and like most propaganda, there was a substantial backstory.

If you were to visit Room 10a of the British Museum, you would find the backstory. An entire room filled with large stone orthostates carved with low relief images of the royal, ritual lion hunt of Ashurbanipal. These stone panels once decorated Ashurbanipal's North Palace in Nineveh. What they depict shows us that the Assyrian kings were not hunting lions in the wild but in a royal hunting park. Wild lions were trapped, held captive, and then transported in crates to the park. On hunting day, armed soldiers with shields and dogs were stationed throughout the facility to keep the lions from escaping or doing any real injury to the king. The desperate courage of the lions, and the wanton slaughter of these majestic creatures, is so graphic and so lifelike that the first time I visited the British Museum I had to leave the room—and I was not the only one. In fact, more than one art historian has theorized based on the staid portrayal of the kings, and the immensely animate and detailed depiction of the lions, that the sympathies of the original artist were with the lions. Why display such graphic images of slaughter? To demonstrate to the Assyrian citizenry, in accordance with their value system, the great valor of their king, and therefore his right to rule.

And, yes, at this point, you might be thinking of another king who did indeed prove his right to rule by killing a lion, alone, with his bare hands, long before his actual coronation. But unlike the Assyrian kings, this young

Figure 5. The lion hunt

hero did not kill for sport. He killed because his father's flock was in danger. This heir to the throne of God's kingdom had no weapons, dogs, or armed guards, or press release for that matter. But that is another story.

Returning to the Neo-Assyrians, one of the wall reliefs in the British Museum depicting the royal hunt caught my eye back in 2010 when I was working on an article for the *Bulletin of Biblical Research*. In this relief (see fig. 6), the Neo-Assyrian king is again returning from the hunt. His right to rule is once again being celebrated by the iconographic display of the now familiar remains of a slain male lion. But what arrested my attention is that in addition to the slain lion is the display of a mother bird *with* her nest still full of eggs.[19] In other words, this relief shows that in Neo-Assyria, taking a mother bird *with* her eggs, a "mother with her children," was worthy of display as another testimony of a sovereign's right to rule.

Figure 6. Assyrian eunuchs return with the prizes of the royal hunt

But in Deuteronomy Israel is commanded to be different: "You shall not take the mother [who is sitting] on the young. Rather, you will surely shoo the mother away, and the young you may take for yourself, in order that it may be well with you and that you may prolong your days" (Deut 22:7). In contrast to the practice of their neighbors, Israel is instructed in the *wisdom* of preserving the creatures with whom they shared the Promised Land. Indeed, Deuteronomy states that if Israel killed off the wild creatures without a thought as to the creatures' ability to replenish their populations, it would *not* "be well" with Israel in the land. I believe the same would apply to us.

CASE STUDY
The Black Bear in the Bottomlands of Mississippi

In February 2019 I was privileged to be invited to the United Methodist Clergy Leadership Conference in Brandon, Mississippi, to offer a three-day curriculum titled "Holiness in a Modern World: An Inquiry into Creation Care." As I had served for four years in an inner-city seminary in Jackson, I shared strong relationships with many people in the room. But even so, I was more than a bit surprised that *this* was the curriculum they wanted, and more than a bit apprehensive as to what the results might be. So, knowing I was speaking to Mississippians, I did some research on endangered indigenous species. Predictably, the primary cause for the loss of native species in the great state of Mississippi was habitat loss. In this case, the clearing and draining of vast acreages of the majestic bottomland hardwood forests in the Mississippi Alluvial Valley (known by the locals as "the Delta").[20] By 1980, more than 80 percent of these forests were gone.[21] Whereas this particular ecosystem had once supported an array of wildlife,[22] the bottomland hardwood forests had been stripped to make way for agriculture—first cotton, then industrial pine production. As a result, species such as the black bear—an animal so prolific and emblematic to the state that President Teddy Roosevelt himself came to hunt black bear in Mississippi[23]—were decimated. As a result, the last known birth of black bear cubs in Mississippi was in 1979. Even so, it wasn't until 1984 that the black bear found its way onto the state's endangered species list.[24]

How did this happen? The first and most fundamental cause was habitat loss. But this principal issue was significantly compounded by unsustainable hunting practices. And in this case, no one sounded the alarm until it was too late. Because the *means* of life had not been protected (habitat), because both mother and offspring had been taken (unsustainable hunting practices), the extermination of a particular species in a particular place had resulted. On the day of my presentation, the local sportsmen and conservationist groups were celebrating the fact that *two* breeding females had been identified in the furthest southwestern county of the state.[25] The reason these reports embody so much hope is that it is females who become resident in a territory, and therefore two lingering females means that there was some chance that black bears were coming home. But what a paltry representation of the thousands of black bears that had once called the Delta home. And as bears reproduce at a relatively slow rate, recovery will be equally slow and frighteningly fragile. In this equation, adult mortalities are devastating—vehicle collisions, electrocutions, poaching, accidental trapping, and hunting could make recovery impossible. Most troubling here is that, as all biologists will tell us, when the top predator in any food chain is removed, the trophic cascade has devastating ramifications at every level of an ecosystem.

So the United Methodist ministers of Mississippi and I grieved this emblematic loss to their homeland—a loss intricately intertwined with the systemic evils of shortsighted leadership, the unjust victories of industrial agriculture, the collapse of the family farm, decimated local economies, and the marginalization of the voiceless. We were encouraged that there are currently efforts from the Wetland Reserve Easement and the Conservation Reserve Program to begin reclaiming the Delta. The hope is that some of the decimated bottomland hardwood forests will be restored, and the isolated patches of surviving habitat might be connected to provide corridors for the black bear and other species like it. But as James Allen documents, forty to fifty years after the first restoration efforts began, the reforested stands cannot compare with the diverse flora of Mississippi's past. Rather, the "near monocultures"[26] reconstructed by

ʾādām cannot compare with the magnificent design of the Creator. Centuries of neglect will take more than a few decades to repair.

So we return to the biblical mandate. The wild creatures that God has placed in our care are not ours, nor are they simply disposable. These creatures and their habitat are vulnerable. And it is *ʾādām*'s God-ordained task to deploy our superior gifting to conserve and protect, not to exploit and abuse "in order that it may be well with you [us] and that you may prolong your days" (Deut 22:7).

DISCUSSION QUESTIONS

1. Why might God be interested in preserving the wild creatures in the ancient land of Israel?

2. Were other nations around Israel practicing such preservation?

3. Do you see value in protecting the wild creatures in your own neighborhood, town, state, country?

4. If you do see the value of protecting the wild animal and its habitat, how might you start acting on that value in your own life?

5. What kind of an impact would it have in your community if your church started working for the preservation of the wild creature?

5

ENVIRONMENTAL TERRORISM

War must be, while we defend our lives against a destroyer who would devour all; but I do not love the bright sword for its sharpness, nor the arrow for its swiftness, nor the warrior for his glory. I love only that which they defend.

FARAMIR, IN J. R. R. TOLKIEN, *THE TWO TOWERS*

WHAT DOES THE BIBLE SAY?

Surely if there was ever an appropriate time to sacrifice the long-term fecundity of the land for short-term gain, it would be in the crisis of warfare. Any warzone from the days of Pharaoh Thutmose III's campaigns in Canaan, to World War I's No Man's Land, to the ongoing legacy of Vietnam's Operation Ranch Hand would testify to this broadly held belief. Yet in Deuteronomy 20:19 we find an interesting little law that seems to speak against this conventional wisdom:

> When you are besieging a city for many days in order to wage war against it to capture it, you shall not destroy its trees by swinging an axe against them. Indeed, you may eat from them, but you shall

not cut [them] down. For is the tree of the field a man that it should be besieged by you? Only a tree that you know does not produce food may you destroy and cut down, then you may build your siege works against the city with which you are at war until it falls.

There is a long tradition of commentary on this verse, all of which recognizes the biblical author's effort, for whatever motivation, to reduce the collateral damage inflicted by siege warfare (cf. 2 Kings 3:19). Like Deuteronomy 22:6-7, many have identified this law as a *pars pro toto* (a part or aspect of something taken as representative of the whole) or an *analogia* (a vehicle of Wisdom literature that formulates a more abstract point by way of a practical example).[1] Again, the common idea between these texts is the preservation of the *means* of life, in other words, sustainability.[2]

The potential damage of siege warfare in the land of Israel is illustrated by the long list of indigenous fruit- and nut-bearing trees in the region, and the significant role these trees played in the Israelite economy and diet. Oded Borowski lists the fig, olive, date, sycamore (*Ficus sycamorus*), apricot, carob, almond, pistachio, and walnut trees, as well as several that cannot be identified with certainty by means of their biblical appellatives.[3] All of these trees faced similar developmental realities—if maintained, they would produce for generations, but full maturity preceded production. How long did full maturity take? The all-important olive tree takes five or six years to *begin* to flower and as many as twenty years to reach full production. Even then, olive trees only bear fruit every other year. Famed Harvard archaeologist Lawrence Stager comments, "It is commonly said that one plants an olive yard not for one's self but for one's grandchildren."[4] Similarly, Assyriologist Steven Cole reports that the female date palm—a treasured source of preservable, calorie-rich fruit—"may take as long as twenty years before [it] produce[s] [its] first fruit."[5] As the great dream of every Israelite citizen was to "live in safety, every man under his vine and his fig tree, from Dan even to Beersheba," these timelines of maturation were significant, and the potential disruption of decades of cultivation due to warfare was terrifying (1 Kings 4:25; cf. 2 Kings 18:31).

In light of the long-term value of food-bearing trees, it is no surprise that a standard aspect of Neo-Assyrian military strategy was the decimation of a besieged enemy's vineyards and orchards (see chapter four for a fuller description of the Neo-Assyrian Empire). The objective of such environmental terrorism was first to intimidate. If fear motivated a city to capitulate, the empire obtained its prize without expending the resources necessary for conquest. If intimidation did not achieve the desired results, the end goal was to cripple the city's economic stability for decades to come, regardless of whether the siege was successful. In this fashion, many potential rebels were held in check. The royal annals make it expressly clear that the Neo-Assyrians communicated this wartime strategy early and often through text and image.[6] Hence, Sargon II boasts regarding his assault on the store city of Ursal:

> I entered triumphantly. . . . Into his pleasant gardens, the adornments of his city which were overflowing with fruit and wine . . . came tumbling down. . . . His great trees, the adornment of his palace, I cut down like millet. . . . The trunks of all those trees which I had cut down I gathered together, heaped them in a pile and burned them with fire.[7]

Figure 7. Sennacherib's troops cutting down date palms outside a southern Mesopotamian town

Regarding his siege of the city of Suhu, Shalmaneser III declares, "We will go and attack the houses of the land of Suhu; we will seize his cities. ... We will cut down their fruit trees."[8]

Steven Cole offers an encyclopedic collection of these texts and images, showing that this particular military strategy was a staple of Assyrian war craft.[9] Not new with the Assyrians, this siege tactic may be traced back into the second millennium BCE among the Babylonians, Hittites, and especially the Egyptians.[10] King Hazael's ruination of the ancient Philistine city of Gath (Tel es-Sâfi) demonstrates that this sort of environmental terrorism continued among the Aramaeans, and the archaeological remains at Sâfi poignantly illustrate the magnitude of *permanent* environmental damage that could result from siege warfare.[11] So we see that the systematic annihilation of orchards and olive yards in order to cripple the life-support systems of the enemy was a staple of ancient Near Eastern warfare before and during Israel's occupation of the Promised Land. Yet the book of Deuteronomy forbids it.

What might be the rationale for Deuteronomy's law? To quote Michael Hasel, Israel is forbidden from such military tactics because "it would not be in Israel's interest to destroy the very resources that would later sustain them."[12] In other words, although environmental terrorism might deliver instant results in the midst of conflict, the long-term detrimental impact on one's own or the enemy's life-support systems was ultimately self-destructive. We do not need to look far in American history to find painful testimony of the same.

CASE STUDY
Operation Ranch Hand

Operation Ranch Hand was part of the American offensive directed against Vietnam, eastern Laos, and parts of Cambodia during what is known in the States as the Vietnam War. Between 1962 and 1971, the United States military sprayed 20 million gallons of chemical herbicides and defoliants mixed with jet fuel over one-quarter of South Vietnam and the bordering areas of Laos and Cambodia.[13] The advertised objective was to reduce American and allied casualties by defoliating forested and rural land, depriving guerrillas of cover. What is less broadly known is that the other objective was to "induce forced draft urbanization"[14]— that is, to destroy agricultural land such that rural farmers were forced to relocate into urban centers. The goal was to strip the North Vietnamese of indigenous support and food supply in the south. Estimates count the destruction of agricultural land at 25 million acres in South Vietnam alone (that is 39,000 square miles). Similar to Neo-Assyrian tactics, this land was stripped of its fertility not only in the moment of crisis but for generations to come. How? The 25 million acres sprayed by the US military were infected with levels of TCDD (Tetrachlorodibenzo-P-dioxin) one hundred times those considered safe by the US EPA.[15] To this day, remaining "hotspots" (such as the Danang military base where Agent Orange was stored, mixed, and loaded onto aircraft) remain at three to four hundred times what is considered safe levels.[16] Clearly this strategy destroyed the "very resources that would later sustain" the South Vietnamese.[17]

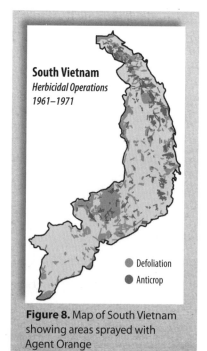

Figure 8. Map of South Vietnam showing areas sprayed with Agent Orange

One of the heartbreaking results of the ongoing concentrations of these deadly chemicals is the thousands of Vietnamese children who have been born with profound birth defects and disfigurement due to prenatal exposure to the dioxins. The images from the Ho Chi Minh City's Go Vap orphanage are nearly impossible to bear.[18] According to the Red Cross of Vietnam, 4.8 million Vietnamese people have been exposed to Agent Orange, and of these four hundred thousand have died from related causes.[19] About one million of those exposed are currently disabled or have related health problems, including cancers, birth defects, skin disorders, autoimmune diseases, liver disorders, psychosocial effects, neurological defects, and gastrointestinal diseases.[20] And since these chemicals are capable of actually damaging genes, it's possible that many generations will continue to suffer the resulting birth defects and deformities from exposure.

And what of the young American patriots who were responsible for dispensing these herbicides? The US Department of Veterans Affairs currently lists Parkinson's disease, heart disease, and cancers of the lung, larynx, trachea, and prostate as "presumptive" diseases associated with exposure to Agent Orange.[21] My own father, a Vietnam veteran with very limited exposure to Agent Orange, died from prostate cancer at the age of sixty-two. These ailments have crippled the lives and earning capacities of tens of thousands of US veterans. In addition to immeasurable emotional pain, the US Department of Veterans Affairs reports that there has been $2.2 billion paid out in retroactive benefits.[22]

Then there are the economic effects. Not only has South Vietnam, our ally, been economically crippled by our utilization of environmental terrorism against our shared enemy, but in 1984 Monsanto and its allied chemical companies were required to pay $180 million in damages in an out-of-court settlement—a settlement bitterly disputed because the compensation offered to victims was absurdly small.[23] This settlement offers an Agent Orange widow $3,400. It offers a veteran with a proven health issue $12,000 over the course of ten years, *if* the veteran is willing to surrender any other state aid.[24] And as for the ongoing hot spots in South Vietnam—areas so contaminated with the residue of these lethal dioxins that they *continue* to create crippling health issues for the citizens of that country—$43 million was committed by the US government in 2009 to begin the now forty-five-year-old cleanup.[25] Unequivocally, Operation Ranch Hand was ultimately self-destructive for ally and enemy alike.

And so we return to the wisdom of the ancients. The abuse of the land was forbidden—whether for the sake of economics or even for national defense. In Israel the human populace was commanded to set their sights on the long-term fecundity of the land. And the fact that it took a generation for an olive orchard to come to full fruition demanded deference. In God's government, human enterprise and aggression simply were not worthy excuses for wiping out the future productivity of the land, the precious ecosystems that inhabited it, or the humans whose lives relied on those systems.

DISCUSSION QUESTIONS

1. Why might God outlaw environmental terrorism in Israel's case?
2. Were other nations around Israel practicing environmental terrorism?
3. Did the fact that Israel was forbidden to practice environmental terrorism put them at risk among their neighbors?
4. How could you apply this Israelite law to your life and participation in society? How might you advise those under your care?

6

THE WIDOW AND THE ORPHAN

Learn to do good;
seek justice.
Reprove the ruthless;
defend the orphan;
plead for the widow.

ISAIAH 1:17

This is pure and undefiled religion in the sight of
our God and Father, to care for orphans and widows
in their difficulties, and to keep oneself
unstained by the world.

JAMES 1:27

THERE IS NO QUESTION that the Bible is on the side of the marginalized. As I discussed in chapter one, God's blueprint for creation was a world in which 'ādām would succeed in directing and harnessing the amazing resources of this planet so that there would always be enough. Progress would not necessitate pollution, expansion would not require extinction, and the privilege of the strong would not demand the deprivation of the weak. Yahweh's world was a world in which there would never be hunger, homelessness, abuse, famine, genocide, or refugee camps. But as a result of the fall, all of these realities became resident on our planet. The ultimate objective of God's great plan of redemption is to fix that.

If you have read or studied my *Epic of Eden* materials, you know that a major advance along the journey toward that final restoration is the Mosaic covenant. God's people (identified through Abraham) finally find rest in God's place (Canaan) with access to God's presence (the tabernacle). Although the formation of the nation of Israel is not the final objective of redemption (we will need the new heaven and new earth for that [Rev 21:1]), the relationship that Israel had with their land serves as a model of God's intentions for our relationship with the land. As Israel understood that their land and its produce ultimately belonged to Yahweh, one of their responsibilities as land stewards was to manage the fruit of the land such that the needs of the marginalized were met. In other words, the farmer was expected to reserve a portion of his harvest for the widow, the orphan, and the "resident alien" of his village.[1]

WHO WERE THE WIDOW AND THE ORPHAN?

Israelite society was very different from contemporary life in the urban West. Whereas modern, urban, Western culture may be classified as "bureaucratic," Israel's society was traditional, most specifically, "tribal."[2] As I have detailed in *The Epic of Eden*, in a tribal society the family is the axis of the community, and an individual is linked into the legal and economic structures of their society through their identity as a member of a particular family. Because Israel's was a patriarchal tribal culture, the linchpin that connected each household to the larger society was the oldest male member of the family. The patriarch was responsible for the economic well-being of his family, he enforced law, and he was responsible for his extended household members who became marginalized through poverty, death, or war. Obviously, this is very different from a bureaucratic society in which the state creates economic opportunity, enforces laws, and cares for the those in need.[3] Ancient Israel had no social security, unemployment benefits, American Disabilities Act, TANF, Medicaid, CHIP, SNAP, EITC, Supplemental Security Income, subsidized housing, or housing assistance. Nor did ancient Israel have a police force, a foster-care system, public hospitals, or orphanages. Rather, in Israel and societies like it, the family cared for individuals who found themselves on the margins. Moreover, a kinsman's responsibility to care for a relative was directly proportional to the proximity of the bloodline between the family members. In other words, the closer the relative, the higher the level of legal and economic responsibility.

In Israel's particular form of patriarchal tribalism, society was formed by a "progressively inclusive series of groups" emanating from the patriarch of the household.[4] Ever broader circles radiated out from the closely knit "father's house," to the clan or "lineage,"[5] then to the tribe, and finally to the nation, as pictured in figure 9.[6] Even the terminology for "family" in ancient Israel reflects the centrality of the patriarch, as the basic household unit was called the "father's house" (Hebrew *bêt ʾāb*). This household included the patriarch, his wife (or wives), his married sons with their wives, his unwed children, and his grandchildren. Current ethnographic studies indicate that the Israelite *bêt ʾāb* could include as many as three generations, fifteen to twenty persons.[7] When a man married, he remained in the household, but when a woman married, she joined the *bêt ʾāb* of her new husband. As a result, both her location and her tribal affiliation shifted to that of her new family, and her children became the possession (and heirs) of her new household. This is why Rebekah leaves Paddan-aram (modern-day Syria) to join Isaac in Canaan (Gen 24), and why Rachel and Leah eventually leave Paddan-aram to follow Jacob to Canaan as well (Gen 31). This "father's house(hold)" lived together in a family compound, collectively farming their inherited land (patrimony), sharing their resources and their fate.[8] Legal decisions involving discipline were enforced through the household, and provision for those impoverished or abandoned came through it as well. Thus those who found themselves without a *bêt ʾāb*—the widow, the orphan, or the resident alien—also found themselves outside the society's normal circle of provision and protection.[9]

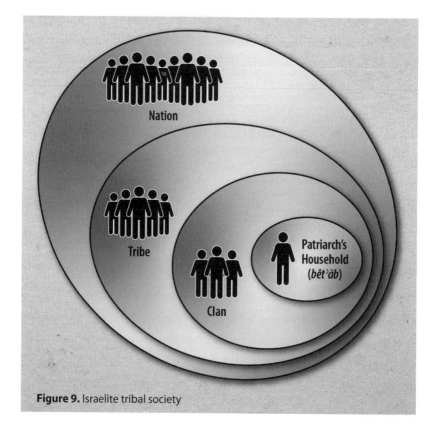

Figure 9. Israelite tribal society

Job offers us a heartfelt description of the fate of the orphan and widow:

> They harvest their fodder in the field,
> and they glean the vineyard of the wicked.
> They spend the night naked, without clothing,
> and have no covering against the cold.
> They are wet with the mountain rains;
> they hug the rock for want of a shelter.
> Others snatch the orphan from the breast,
> and against the poor they take a pledge.
> They cause [the poor] to go about naked without clothing,
> and they take away the sheaves from the hungry. (Job 24:6-10)

Because of her lack of an advocate, a widow could indeed have her child "snatched from her breast" to pay her debts. And without a household to provide for them, orphans could indeed wander about homeless, cold, and naked. Hence, in his defense against his accusers in Job 31, Job specifically reports how he has cared for the widow and the orphan as testimony to his good character.

> If I have kept the poor from their desire,
> or have allowed the eyes of the widow to fail,
> if I have eaten my morsel alone,
> and the orphan has not shared it [then I would be guilty]!
> But from my youth [the orphan] grew up with me as with
>   a father,
> and from infancy I guided her.
> If I had seen anyone perishing for lack of clothing,
> or that there was no covering for the needy,
> if his loins have not blessed me because he has been warmed with
>   the fleece of my sheep,
> if I have lifted up my hand against the orphan,
> because I saw I had support in the gate,

then let my shoulder be dislocated,
and my arm be broken off at the elbow. (Job 31:16-22)

Clearly, this wealthy householder has not only taken care of his own family members but has also shared his wealth with those less fortunate. The evidence of his integrity is his quantifiable acts of charity to those *outside* his household.

Because Israel's tribal culture was patrilineal—meaning that ancestral descent was traced through the male line—a woman's identity and her link to the economic and civil structures of her community were always identified through the men in her life. She was first her father's daughter, then her husband's wife, and then her son's mother. The resources and protection of the clan came to her through the male members of her family, who were the only ones in this society who could inherit property under normal circumstances. This is why it was critical for a woman to marry and to bear children. A woman who was widowed prior to bearing a son was a woman in crisis. And a woman without father, husband, or son was destitute. An orphan—that highly unfortunate child who had no remaining relatives to care for her or him—was in worse straits. Because of these societal realities, there were a number of laws in Israelite society that addressed the protection of the widow and the orphan.

Consider, for example, the *levirate law*. Found in Deuteronomy 25:5-10, this law protected the young widow and preserved proper lines of inheritance for her deceased husband. The Latin term *levir* means "brother," and the law dictates the behavior of surviving brothers when a man has left a young and childless widow behind. In such instances when a *bêt ʾāb* had more than one adult son, the premature death of a young husband required his closest remaining brother to marry his widow. The objective was to produce a male heir for the deceased and to keep the young widow within the protective walls of the father's household. Thus the first child of a levirate union was legally recognized as the deceased's heir, and any additional children belonged to the living brother. The intent of this law was both to protect the young widow from destitution and to protect her deceased husband's inheritance. As the story of Judah, Tamar, Er, Onan, and Shelah in Genesis 38 illustrates, the people of Israel considered it a very serious offense for a man to fail to fulfill this responsibility to his dead brother, and even more serious to leave a young widow destitute. As dictated in Deuteronomy, a brother who failed to care for his brother's widow would be shamed by his community.

> When brothers live together and one of them dies and has no son, the wife of the deceased shall not be married outside the family to a strange man. Her husband's brother shall go in to her and take her to himself as wife and perform the duty of a husband's brother to her. And it shall be that the firstborn whom she bears shall assume the name of his dead brother, that his name may not be blotted out from Israel. But if the man does not desire to take his brother's wife, then his brother's wife shall go up to the gate to the elders and say, "My husband's brother refuses to establish a name for his brother in Israel; he is not willing to perform the duty of a husband's brother to me." Then the elders of his city shall summon him and speak to him. And if he persists and says, "I do not desire to take her," then his brother's wife shall come to him in the sight of the elders, and pull his sandal off his foot and spit in his face; and she shall declare, "Thus it is done to the man who does not build up his brother's house." And in Israel his name shall be called, "The house of him whose sandal is removed." (Deut 25:5-10)

Although this system seems very odd to most Westerners, it worked. The inheritance of the deceased brother was properly conferred on his legal offspring, and the young widow was secured within the *bêt ʾāb* of her deceased husband. As a result, the widow's need for food and shelter was met, and her future need for a child to care for her in her old age was addressed as well.[10]

The book of Ruth offers us a beautiful memoir of what life on the margins looked like for the widow and orphan during Israel's settlement period. As the story begins, we see that Elimelech was a fairly typical "small holder" farmer, "barely making it" on his ancestral land in the village of Bethlehem.[11] Based on our archaeological reconstructions of the era, Elimelech likely was practicing dry farming (rain-fed agriculture), growing grain, grapes, and olives while keeping a small family flock for meat, dairy, and textiles. Bethlehem was a small town, apparently without a defense wall, but likely safeguarded by encircled domestic dwellings such that the exterior walls of houses shielded the village from harm.[12] But as is typical in the hill country, there was a drought. And the drought led to a famine. And Elimelech was left with no seed for his next planting and no choice but to abandon his patrimony and try to make a go of it with his wife and young sons across the river in Moab. Although Moabites were the notorious "other" in the Israelite experience (in other words, the people we *don't* like[13]), the plan worked. The family found land, settled for ten years, and was able to earn both the confidence of the locals and the necessary means to negotiate two marriages for their sons. But before we reach Ruth 1:5 tragedy strikes—Elimelech dies, and so do Naomi's two adult sons. In a patriarchal tribal culture this family has just become an unfamily. These three women are in serious trouble. There are no males left in the household to marry the young widows. Naomi is living in a foreign land far from Elimelech's patrimony and the protection of his clan. There are no legally bound relatives to whom they can turn for aid. So Naomi begs her daughters-in-law to return to their households of origin, as she has no means by which to provide for them (Ruth 1:8). And this widow-of-widows decides to leave alone for Bethlehem seeking the safety of her extended kinship circle. Naomi's hope is that the girls' mothers' affection for them will override the limited resources and legal realities of the day, and these girls will be remarried to Moabite men (Ruth 1:11-13). Ruth's excellence is demonstrated here in her commitment to her mother-in-law, which surpasses any legal or societal expectation. In Ruth 1:16-18 Ruth utilizes every possible expression of kinship alliance to announce to anyone who will listen that she is not going anywhere. Even though the linchpins of their lives are gone (their husbands and for Ruth her father-in-law), Ruth has declared Naomi her kin, with all of the responsibilities and privileges that come with that title. I can assure you that for the original Israelite audience, the drama of this story was most simply how in the world these two widows were going to find safe space.

The answer to this question will come via the exceptional integrity of the other lead character, Boaz. His remarkable commitment to the widows (one of whom might also be identified as an orphan) becomes the gold standard of the biblical text. Boaz welcomes the outsider, he looks past her ethnic and socioeconomic differences, and he *redeems* both Ruth and Naomi even though there were other kinsmen who were closer in bloodline and should have stepped up to the plate. Boaz resolves the life-threatening realities of Ruth and Naomi's situation by buying back their land, finding a place in his household for them, fathering a child in his deceased kinsman's name, ensuring the child's inheritance, and, of course, becoming the grandfather of Jesse, the father of David—the father of Jesus (Ruth 4:16-22; Mt 1:1). Both Ruth and Boaz demonstrate that they are people of *ḥayil* ("excellence" and "strength"; Ruth 2:1; 3:11; 4:11) by their dogged commitment to the *widow* Naomi's well-being.

In sum, anyone finding themselves outside the *bêt ʾāb* in Israel's world was in dire circumstances. Therefore, much of Israelite law was designed to protect the *bêt ʾāb* such that land tenure was ensured and widows and orphans were not created via the chaotic impact of premature death. The ideal was that no one would fall outside the sheltering walls of the father's household.[14] But when the worst happened, charity was required.

As we learned in chapter three ("The Domestic Creatures Entrusted to *ʾĀdām*"), throughout its national period, most of the Israelite populace lived on what Carol Meyers has dubbed "small-holder family farms." This means that the common man in Iron Age Israel, like Elimelech in the book of Ruth, lived the fragile existence of a subsistence farmer.[15] As we have reviewed, in the central hill country the main economy was a mixture of pastoralism and intensive, permanent, diversified agriculture. Even in a good year, the typical *bêt 'āb* could expect a "hungry season" of sixty days or more. We can well imagine the pressure under which the responsible patriarch lived his life—fifteen to twenty people to house, feed, and clothe; the hostile conditions native to dry farming; and the constant threat of drought, disease, and war. One short harvest could place his entire family in harm's way. Yet even under these conditions, when we again turn to the constitution and bylaws of the nation (the book of Deuteronomy) we once again find a radical call to discipleship:

> When you reap your harvest in your field and have forgotten a sheaf[16] in the field, do not go back to get it; let it be for the resident alien, for the orphan, and for the widow, in order that Yahweh your God may bless you in all the work of your hands. (Deut 24:19; cf. Lev 19:9; 23:22)

As we have learned, grain (wheat and barley) was the backbone of Israel's domestic food supply. This is how the Israelite farmer kept both his children and his livestock fed. It became a critical commodity in international trade as their economy advanced.[17] Yet our subsistence farmer is being instructed to refrain from fully harvesting his most essential dietary anchor. A similar command is offered regarding his olive yard: "When you beat your olive tree, do not go over the boughs again; let it [the unharvested portion] be for the resident alien, the orphan, and the widow" (Deut 24:20; cf. Lev 23:22). Like grain, the olive was fundamental to ancient Israel's economy. Its oil was not only indispensable to domestic survival but had also long served Canaan as a significant export—a "cash crop" of sorts (see 1 Sam 8:14; 1 Kings 5:11 NIV [1 Kings 5:25 NAB];

Hos 12:12; 1 Chron 27:28). As Lawrence Stager summarizes, "The production of olive oil was a major industry, accounting for much of the economic prosperity of the region. Surplus oil was exported to Egypt, Phoenicia, and perhaps even to Greece."[18]

Viticulture (the cultivation of grapes) was the third building block of the domestic and commercial venues of Israel's economy. Canaan was (and is) famous for its wine, and grape production thrived in this region as far back as the Early Bronze Age.[19] In fact, the wine was so famous that Pharaoh Thutmose III's Karnak botanical garden depicts grapevines imported from Canaan to Egypt—the Egyptians' attempt to import Canaan's expertise into their world.[20] Yet Deuteronomy commands that the gleanings of the vineyard (Hebrew: ʿōlēlôt) be left for the poor. Leviticus further particularizes this command stating that the smaller clusters (Hebrew: *peret*) be left as well.[21]

> When you gather the grapes of your vineyard, do not glean afterward; let it [the unharvested portion] be for the resident alien, the orphan, and the widow. And remember that you were a slave in the land of Egypt; therefore I am commanding you to do this thing. (Deut 24:21-22; cf. Lev 19:10; 23:22)

Thus the law code of Deuteronomy lists the three crops essential to Israel's agricultural cycle (and therefore survival), and demands that the small-holder farmer *not* fully harvest his crops. Recognizing the subsistence struggles of the typical family farm, this is a big ask.[22] What principle do we the modern readers find here? God's command was (and is) that the produce of the land be shared with the widow, the orphan, and the resident alien so they might have the opportunity to sustain themselves. In other words, the drive for economic security and surplus in Israel must always be tempered by God's command for charity. Not even economic viability served as an acceptable rationale for greed. In sum, the Israelite citizen was instructed that the land did not belong to him; it belonged to God. And God wanted the marginalized to have the chance to benefit from its produce too.

WHAT DO WE SAY?

What most of us do not realize is that environmental degradation strikes those on the margins first. It is the subsistence farmer and the poor who pay the highest price for any society's failure to utilize land in a sustainable fashion. As Norman Wirzba, research professor of theology, ecology, and rural life at Duke Divinity School, discusses at length, history is filled with examples of shortsighted agricultural practices that turned fertile fields into wasteland and desert, thereby effecting the collapse of civilization in particular regions at particular times.[23] As discussed above, in the field of Old Testament studies, the most familiar story is that of Mesopotamia's agricultural collapse due to a failure to fallow (see the section in chapter two "Sustainable Agriculture: What Does the Bible Say?").[24] Wendell Berry's classic treatment of this topic in *The Unsettling of America: Culture and Agriculture* (1977) is a more contemporary example. David Montgomery's *Dirt: The Erosion of Civilization* (2007) offers a pointed discussion of the same. Each of these describe the all-too-human cycle of exploitation, abandonment, and reexploitation of cultivatable soil.[25] In our world, however, "the problem . . . is that we have run out of places to exploit. The frontier is closed. The question we now face is, how are we going to live sustainably . . . where we *are*? Can we grow food in ways that do not imperil the ability of future generations to feed themselves?"[26] As Oaxaca, Mexico, and Port-au-Prince, Haiti, painfully illustrate, eroded and desiccated farmland equals poverty, starvation, and mass migration.[27]

The exploitation of the land, however, is not limited to agriculture. Deforestation occurs for many reasons, and people such as Scott Sabin (executive director of Plant with Purpose, an international Christian organization that empowers the marginalized impoverished by deforestation) have spent their lives attempting to make the public aware of the relationship between shortsighted environmental abuse and refugee populations.[28] Pavan Sukhdev, author of The Economics of Ecosystems and Biodiversity (TEEB) report,[29] states:

> Poverty and the loss of biodiversity are inextricably linked: the real beneficiaries of many of the services of ecosystems and biodiversity are predominantly the poor. The livelihoods most affected are subsistence farming, animal husbandry, fishing and informal forestry—most of the world's poor are dependent on them.[30]

In this report Sukhdev argues that the "ecosystem services" on which the poor depend are predominantly public goods. But as there is no recognizable market or pricing for these commodities, "they often are not detected by our current economic compass."[31] Thus, when the pressures of expanding industry, population growth, changing diets, urbanization, and climate change undermine healthy, diverse ecosystems, the well-being of the poor is the first to be compromised. But because no one *owns* the natural systems, these losses don't show up on anyone's financial or economic spread sheet—and the effects of environmental degradation on the marginalized remain invisible.[32]

One example is the republic of Haiti. Centuries of unsustainable agricultural methods and inefficient charcoal production have created a staggering level of deforestation that has been both the result and the cause of intolerable conditions for the widow and the orphan.[33] Aerial photographs of the border between Haiti and its island companion, the Dominican Republic, show a verdant rainforest (the DR) bordered by a dirt road, beyond which is a completely denuded landscape (Haiti). With an average annual income less than $400,[34] and a capital city (Port-au-Prince) ranked fourth out of the 230 worst cities on the planet to inhabit, the living conditions in Haiti are beyond challenging.[35] But for the poor, they are unbearable. Matthew Ayers, president of Emmaus University just outside Cap-Haitian, has served this community for over a decade.[36] Matt has too many stories to tell about the incredible challenges that the locals endure on a daily basis, as well as the fantastic hurdles that he and his staff have had to overcome to keep Emmaus up and running. Ayers speaks of forced urbanization due to the deforestation of the countryside, and how the resulting overflow of humanity in the cities paralyzes the little infrastructure that Haiti has. Thus the disenfranchised are driven to the least desirable areas of the city—areas adjacent to the rivers and ocean. Living near the water in Haiti's overpopulated cities is particularly dangerous because deforestation "means frequent flashfloods and landslides that are almost always fatal."[37] Add to this the array of diseases resulting from miles of open sewage and standing water (e.g., malaria), and the relationship between environmental degradation and the plight of the widow and the orphan is brought into sharp focus. In Haiti deforestation means forced urbanization, and forced urbanization means the victimization of the poorest of the poor.

Neal and Danielle Carlstrom, who serve as World Venture missionaries to Madagascar, tell a similar tale about the lush forests and fertile waterways that once graced the island. Ninety percent of the flora and fauna in Madagascar is endemic—truly extraordinary and unique. This island has extravagant natural resources. But predatory exploitation has left Madagascar *85 percent deforested*—a statistic I can hardly wrap my brain around. The result? Desiccated topsoil, massive erosion, clogged rivers, and suffocated marine life and coral reefs. The Edenic "Red Island" is now one of the ten poorest countries in the world, and the Malagasy are starving. Neal states that "in severe poverty, you see brokenness, corruption, and evil on all levels but those who suffer the most are always the least of these: the outcasts, the disabled, the widows, the orphans, the weak, the uneducated, the indentured servants, the voiceless."[38] So Neal spends his days teaching the least of these how to create their own microbusinesses by successfully planting and nurturing indigenous trees in their backyard gardens—a beautiful and integrated enterprise that lifts the Malagasy out of unimaginable poverty while empowering them to reclaim their homeland. "When you love, empower, and teach the poorest of the poor how to restore their land and lives, they find hope, stability and a future."[39] Danielle is a midwife. She reports that *one out of ten* Malagasy women die in childbirth and that this absurdly high infant mortality rate is due mostly to malnutrition.

> As a result of the degradation of their land, many of the women we serve are horribly malnourished. This makes them more likely to hemorrhage after labor, makes their babies more susceptible to being born early, small and without the stores they need to thrive. ... Farming families whose land no longer provides struggle to get any sort of food let alone sending their children to school. And so the cycle continues and worsens.[40]

So Danielle spends her days teaching these women how to care for themselves so that their babies might have a chance at life, while Neal teaches their menfolk how to restore their farms and waterways so that life and hope might return to a country plagued with darkness and despair.

In sum, environmental degradation in Haiti and Madagascar has struck the most vulnerable the hardest. Whereas the wealthy have either profited or been protected from the islands' demise, the rural farmer, the widow, and the orphan have been crushed by it.

CASE STUDY
Mountaintop Removal Coal Mining

What of sustainable land use that benefits the widow and the orphan beyond agriculture? Let's pause to consider MTR-VF coal mining. Mountaintop removal (MTR) is a relatively new form of coal mining that requires the targeted site to be clear cut and then leveled by the use of massive amounts of explosives in order to reach the coal seams buried deep within the mountain. Valley fill (VF) emanates from the need to dump the "overburden" created by this method once the coal is extracted. This overburden is the remains (vegetation, topsoil, rock, etc.) of what was once a thriving mountain ecosystem dumped into an adjoining valley. As with most coal mining in our country, the regions targeted for exploitation are West Virginia, eastern Kentucky, southeastern Tennessee, and parts of western Virginia. Although coal mining in the United States has a long and troubled history, MTR-VF has brought the abuse of the land, the miner, and his community to an entirely new level.

Figure 10. A dragline excavator

With MTR, millions of pounds of explosives are used to access what remains of America's coal seams. Demolition may extend as far as one thousand feet below the surface. After demolition, coal and debris are collected by enormous earth-moving machines known as dragline excavators—among the largest machines on earth (fig. 10). The "draglines" required to move the overburden stand twenty-two stories high and typically weigh 8,000 tons, and their buckets have a holding capacity equal to twenty-four compact cars.[41] The largest ever built in the States, the "Big Muskie," is memorialized at the Miners' Memorial Park in southeastern Ohio—the "Big Muskie" weighed in at 12,000 tons and had the largest-capacity bucket ever built.[42]

The overburden excavated in order to reach coal seems is dumped in adjacent valleys. Erik Reece records 6,700 "valley fills" approved in central Appalachia between 1985 and 2001, resulting in over 700 miles of healthy streams buried completely and thousands more damaged.[43] A more recent statistic records 2,000 miles of waterway in the US either poisoned or eradicated by MTR-VF. Can this damage be repaired? A 2010 study conducted by an independent research team of university biologists and geologists states: "Current mitigation strategies are meant to compensate for lost stream habitat and functions but do not; water-quality degradation caused by mining activities is neither prevented nor corrected during reclamation or mitigation."[44] Between 1985 and 2015, explosives and mining equipment chewed up an additional eleven hundred square miles of the Appalachian mountains, but ironically, MTR-VF is only obtaining a third of the coal from the same quantity of stripped land area as it did three decades ago.[45] In other words, the massively destructive methods of MTR-VF are producing diminishing returns.

The demolition required for MTR leaves in its wake what many have described as a "lunar landscape"—where once verdant forests grew, now nothing grows (see fig. 11). The method causes *"permanent* loss of ecosystems that play critical roles in ecological processes," and the impacts on both land and waterways "are pervasive and irreversible."[46]

The impact on the local populations? Activist groups such as Appalachian Voices continue to labor to bring the issue into view. But because MTR-VF has targeted eastern Kentucky and West Virginia, some of the poorest communities in America, few are listening. The locals are hungry for jobs and often support the very same industry that is exploiting them.

Figure 11. Lunar landscapes

Some resist. But those who do are threatened and harassed into selling their land or live with the consequences of ongoing explosions within yards of their homes. Groundwater is poisoned, houses damaged, and lungs filled with the voluminous debris MTR-VF creates.[47]

One such story comes from Rawl, West Virginia. Carmelita and Ernie Brown have lived out their married lives in an attractive brick home just downstream from a mountaintop removal site operated by Massey Energy—what had been the biggest coal company in the state.[48] Ernie was a coal miner, as were his father and his grandfather before him. He and Carmelita and their larger community understood that their livelihood depended on the coal industry, and they had accepted the realities of that relationship. That was, until they started getting sick. And their neighbors started getting sick. Undiagnosable ailments, rashes, cancers, and kidney stones began occurring in people all over Rawl. And then the well water coming from Ernie and Carmelita's tap began to smell of sulfur . . . and turned brown.

Some attempted to compensate by using bleach in their dishwater and laundry. No one realized that what was happening was the well-documented result of mountaintop removal coal mining: the water table from which the entire community drew its water had been poisoned. Arsenic, manganese, lead, barium, selenium, aluminum, and other toxins had leached into their ground water and thereby into their drinking water.[49] The toxins in their water were traced to the mountaintop-removal operation just above the Browns' house.[50] With the help of the media and outside organizations, Rawl shared their plight with the public. After twelve long years, they won a legal battle that resulted in a city water line that replaced the old wells. This resolved their immediate problem, but what about the groundwater? What about local streams, fish, and wildlife? What about the diseases that could not be reversed, and what about the next community poisoned by the same irresponsible mining methods?

One particularly painful story regarding the impact of this industry on local residents involves a mountaintop-removal site in Appalachia, Virginia. In the wee hours of the morning on August 20, 2004, a boulder weighing half a ton was pushed off a blasting site by a bulldozer in a clandestine attempt to widen a supply road. The boulder came crashing down the mountainside, tearing through the side of a young family's home. "It hurled like a cannonball into Dennis and Cindy Davidson's house, right through the wall of the bedroom and onto the bed where 3-year-old Jeremy was sleeping."[51] The boulder stopped just short of seven-year-old Zachary's bed. But three-year-old Jeremy was crushed to death in his sleep.

Flash flooding and potential coal impoundment ruptures are now a constant risk in Appalachia due to the massive changes MTR-VF has made to the local topography. When one such impoundment, again belonging to Massey Energy, failed in Martin County, Kentucky, on October 11, 2000, more than *300 million gallons* of toxic sludge were dumped into the Ohio River, as well as the Big Sandy River and its tributaries. The disaster—nearly thirty times larger than the *Exxon Valdez* spill—killed virtually *all* aquatic life for seventy miles downstream. The EPA named it the worst environmental disaster ever to occur east of the Mississippi. But until I began writing this book, I'd never heard of the event. I was living in central Kentucky on October 11, 2000, and I never saw a single press release. And I've yet to meet anyone who has. The disaster didn't find its way into my newspaper or my daily newsreel—or apparently anyone else's either. Why? Probably the same reason I never heard about the children of Marsh Fork Elementary school in Sundial, West Virginia, the children who spent seven years learning to read, write, and do arithmetic four hundred yards downslope from Massey Energy's Shumate impoundment. This impoundment holds 2.8 billion gallons of toxic coal sludge. As the Martin County impoundment disaster demonstrated, these retention "ponds" are vulnerable to failure. So the plan in case of failure was to sound a bullhorn, allowing the Marsh Fork Elementary School's 230 children five minutes to evacuate before six feet of toxic sludge engulfed their school. Seven years of protest finally resulted in action from Massey Energy—the construction of a new school three miles away from the impoundment. The school was built in December 2012, and now the locals, who are struggling with limited resources in every aspect of their lives, are content. They have a new school for their children that is now located out of harm's way. But should we be content?

The rationale for MTR-VF is, of course, profit. When asked why surface mining is permitted near residential neighborhoods in the wake of the Davidson case, agency spokesman Mike Abbott replied, "Because state and federal laws allow it." As more than one-third of the coal burned in the United States is mined in central Appalachia, and nearly half of the electricity used by Americans is powered by coal, this "cheap" energy source either makes or saves a lot of people a lot of money. But, by the laws of Deuteronomy, should "cheap" or "convenient" be the deciding factor for the community of faith?

The next question I am forced to ask is, What is the church doing about this? Allen Johnson of Christians for the Mountains—a group that describes itself as "a network of persons advocating that Christians and their churches recognize their God-given responsibility to live compatibly, sustainably, and gratefully joyous upon this God's earth"—is desperately trying to make a difference.[52] In the fall of 2015, my coteacher at Wheaton College, Kristen Page, and I invited Allen into our classroom via Skype. The students were well-prepared and eager for the conversation—wondering, as we all were, where the outrage regarding this seemingly abusive and primitive exploitation of the amazing Appalachian mountains might be found. After a very engaging conversation, one of our (wonderful, committed, and idealistic) Wheaton students asked Allen what the church in the region was doing to speak out for both the marginalized and the planet. Allen paused. He diverted. You could even say he dodged. Knowing him to be a man of great integrity, I was surprised. That is, until I received this letter from him the next day.

Sandra,

When reflecting upon our Skype dialogue with your class on October 12, I've thought especially about a question that one of your students raised; a question that invariably I'm asked by outside reporters and other interested people. The question generally takes a shape similar to this. "What are some things that local church congregations are doing about the problem of mountaintop removal?"

It's a question that makes me squirm, because the straight-forward answer is, "Nothing." Which is embarrassing. I usually bring up some "exceptions," but these are rare cases. And then I go into a brief analysis of why congregations are uncomfortable taking on any hot-button local issue that potentially can divide a congregation, get a minister fired, result in a drop of financial contributions, create a membership exodus to other churches. . . .

When Christians for the Mountains was forming during a weekend in May 2005, I took a group of a couple dozen Christians up to Kayford Mountain to meet Larry Gibson who would show us the surrounding mountaintop removal from his "island in the sky."[53] We had told him our group would meet him about noon. As it turned out, we got there about 1:30 pm as our group dawdled and lingered on some sites on the way up. After we had gathered together, Gibson lashed into me, "Why are you so late??!!!" I blustered an apology for our group's tardiness. "No, that's not what I mean," snorted Gibson. "Why are you church people so late to doing anything about mountaintop removal?"

Sandra, there are numerous denominational statements against mountaintop removal. The West Virginia Council of Churches has had a strong statement against mountaintop removal for many years. A number of well-known Christian leaders have come out against the practice. And a number of congregations outside our region have given solid support. My own congregation, 70 miles east of the coalfields, backs my work. But within the coalfields, congregations are typically hush on mountaintop removal.

I often think of the civil rights movement of the 50s and 60s. Southern states were not going to enact legislation to dismantle segregation and other disparities. African-American communities were either resigned to the situation or too fearful to step out for change. But then, a few black leaders rose up, some courageous students stood up, some congregations stirred, they suffered, but then a national movement birthed.

Sandra, I would encourage your class to work on your student's question within a broader context. What are local issues that are *taboo or otherwise risky for a congregation to take on? Why? How do jobs, economic matters, loyalties, culture, or our own privilege and comfort from those services play into the silence of a congregation? How can the prophetic word for local issues come from the pulpit, or will it have to be through individual members within a congregation or through para-church ministries? Do policymakers know that denominational statements are only worth as much as the congregations that back them?*

Thank you, again, for the privilege of interacting with your class. May God bless each and all of you.

Allen

When I read this letter I think about the cost of real change. I think about the price real people pay for speaking out against status quo. Having lived on this planet more years than I care to admit, I know that courage and persecution look way different up close than they do in a TED Talk or in a movie. But when I read this letter I also hear the voice of Isaiah:

> Learn to do good;
> Seek justice,
> Reprove the ruthless,
> Defend the orphan,
> Plead for the widow. (Is 1:17 NASB)

The prophets ask us again and again, "Who will defend the voiceless? Who will speak for the orphan and the widow?" Who will speak up for Carmelita and Ernie Brown, Jeremy Davidson, and the children of Marsh Fork? How will we answer when the Creator asks us where we were when the widow and the orphan and the resident alien were stripped of their homes, their health, and their livelihoods because we chose silence?

DISCUSSION QUESTIONS

1. How do the biblical laws of caring for the widow and the orphan inform us of God's intentions for our relationship with the marginalized?

2. If you were an Israelite farmer, how would you have felt about leaving a portion of your hard-won harvest in the field for the widow, orphan, and resident alien of your community?

3. Who do you think the widow, orphan, or resident alien of our day might be?

4. What is your reaction to the impact that environmental degradation has on the marginalized?

5. Why do you think our churches, our country, and our government are ignoring mountaintop removal coal mining right here in our own backyard?

7

THE PEOPLE OF THE NEW COVENANT AND OUR LANDLORD

What is left at the end of all things? Did Jesus die for plants?
No. Did Jesus die for animals? No. Jesus died for people.
And when it is all said and done, the only
thing that will be left is the church.

Sermon heard in Wheaton, Illinois, July 2016

BOTH CHURCH HISTORY and our own believing hearts tell us that there was something amazing about the laser-like focus of the early evangelical movement. John and Charles Wesley, George Whitefield, and Jonathan Edwards were the founding fathers of the "cross-pollinating revivalistic and evangelistic atmosphere of Britain and North America in the 1730s" also known as Great Awakening.[1] Thousands of churchgoers were reawakened in their faith; thousands more were converted for the first time. The Methodist Episcopal Church of the United States, birthed through the preaching of the Wesleys and Francis Asbury, "went viral" by the turn of the century, claiming four million adherents at its zenith.[2]

Foundational to this evangelical revival was the "different twist" that Riley Case speaks of—the belief that the primary task of the church was to preach the gospel such that humanity might be reconciled with God through the work of the Holy Spirit—the conversion of souls.[3]

> When Methodism jumped to America after the Revolutionary War, Wesley's admonition "You have nothing to do but save souls" was understood by the first American Methodists to mean that salvation from the consequences of sin is available by faith in the blood of Jesus through the New Birth. And with the New Birth comes a changed life through the power of the Holy Spirit. This was a different twist from what anyone else in America had been preaching up to that point and it took the new nation by storm.... As for the Methodists, while only 10% of Americans claimed church membership after the Revolutionary War, by 1850 the percentage was nearly 40% and of that 40% Methodists would claim a full one-third.[4]

But as I stated in the introduction, somehow or another this wholly good and orthodox emphasis on the conversion of souls also resulted in the church's sense that converting souls is the *only* task of the Christian. And therefore any *other* task (such as environmental stewardship) is a distraction from the most essential aspect of our calling. There are many divergent trails of church history and theology that we could pursue at this point to decipher *where* that idea came from. Some would blame it on the ancient heresy of Gnosticism, a form of Greek dualistic thought, which supposedly infiltrated the New Testament to teach us that all matter is evil, and only the nonmaterial spirit realm is good and worthy of our investment.[5] Some would blame it on dispensational premillennialism, which holds that the eschaton will be inaugurated by a sudden and cataclysmic event bringing about the annihilation of the created order, a "sharp break from conditions as we now find them."[6] Some would blame it on the great "liberal/fundamentalist" divide in the 1920s, which left social concerns to the liberals and the conversion of souls to the fundamentalists. Others, such as Lynn White (a twentieth-century American medieval historian), blame our current environmental crisis squarely on the Judeo-Christian ethic, which supposedly posits a dichotomy between people and nature in which "man and nature are two things, and man is master, and therefore, whereas the exploitation of people would be ethically evil, the exploitation of creation was right and good."[7] This alleged biblical perspective has frequently been set in contrast to the supposed more eco-friendly views of other religions, leaving Christianity a villain on the world environmental stage.[8] The charge is that the Bible *desacralized* nature by eliminating polytheism and animism, *subjugated* the created order by giving it to humanity to rule, and *degraded* it by the separation of spirit and matter.[9] According to White, the Judeo-Christian tradition in the Western world is "the most anthropocentric religion the world has seen."[10] All of these avenues could be pursued with benefit.[11] But as the purpose of this book is a biblical theology of humanity's responsibility toward the garden, and I believe we've already demonstrated that the Old Testament is deeply committed to the responsible stewardship of land and creature, let's turn to the New Testament. We will start by examining the New Testament passages that seem to say that the created order will be annihilated at the second coming of Christ, passages that seem to infer that it is right and good to use natural resources as aggressively as possible to pursue the true *telos* (Greek: "the last part of a process"[12]) of the new covenant.

WHAT DOES THE BIBLE SAY?

There are several New Testament passages that are often cited as proof that God's ultimate plan is to dispose of the current planet. The first of these is 2 Peter 3:10-13:

> But the day of the Lord will come like a thief, in which the heavens will pass away with a roar and the elements will be destroyed with intense heat, and the earth and its works will be burned up.

> Since all these things are to be destroyed in this way, what sort of people ought you to be in holy conduct and godliness, looking for and hastening the coming of the day of God, because of which the heavens will be destroyed by burning, and the elements will melt with intense heat! But according to His promise we are looking for new heavens and a new earth, in which righteousness dwells. (NASB)

A second passage is 1 Thessalonians 5:2-3:

> For you yourselves know full well that the day of the Lord will come just like a thief in the night. While they are saying, "Peace and safety!" then destruction will come upon them suddenly like birth pangs upon a woman with child, and they will not escape. (NASB)

A third passage is Revelation 6:12-14, 17:

> I looked when He broke the sixth seal, and there was a great earthquake; and the sun became black as sackcloth *made* of hair, and the whole moon became like blood; and the stars of the sky fell to the earth, as a fig tree casts its unripe figs when shaken by a great wind. The sky was split apart like a scroll when it is rolled up, and every mountain and island were moved out of their places.... "For the great day of their wrath has come, and who is able to stand?" (NASB)

A final passage is Revelation 21:1:

> Then I saw a new heaven and a new earth; for the first heaven and the first earth passed away, and there is no longer any sea. (NASB)

How are we to read these passages? Are the heavens and the earth, the waters above and the waters below, the sea and the dry land with all their flora and fauna truly to be obliterated at the end of the age? Is God's ultimate plan to destroy the garden that he commanded both Adam and Israel to tend and protect? Let's first attend to a concept that lies behind all of these passages—"the day of the Lord."

THE DAY OF THE LORD

The day of the Lord (literally, the "day of Yahweh" [Hebrew: *yôm Yhwh*]) is a concept that reaches back to the very beginning of the biblical narrative. Meredith G. Kline, the great Reformed biblical theologian, identifies the first occurrence of the day of Yahweh in Genesis 3:8, when the Creator enters the garden "in the Spirit of the Day" and the thunderous sound of his entry drives Adam and Eve into hiding.[13] Although often translated similarly to the NLT, "When the cool evening breezes were blowing, the man and his wife heard the LORD God walking in the garden," this passage is more likely a reference to the first instance of divine judgment on our planet. Judgment is delivered first to the serpent for his deception, then to the woman for her foolishness, and then to the man for his treason. But as is always the case, the day of Yahweh also brings mercy and hope:

> And I will put enmity (hostility)
> Between you [the serpent] and the woman
> And between your seed and her seed.
> He [the woman's seed] will bruise you on the head,
> But you shall bruise him on the heel. (Gen 3:15)

There is an image that I believe perfectly illustrates the moment rehearsed above. It was drawn by Sister Grace Remington, Order of Cistercians of the Strict Observance, from the Sisters of the Mississippi Abbey in Dubuque, Iowa. The image offers us Eve and Mary at the same age. One is shamed and grieving, with a serpent wrapped around her leg; the other is pregnant and hopeful, with a hand of comfort on her sister's shoulder. One holds a bitten apple. The other has her bare foot squarely placed on the head of a dying snake. Both have fixed their gaze on the child to come. I love this image. Partly because it is beautiful, but partly because in many ways it captures the nature of "the day."

The *yôm Yhwh* is indeed a day of judgment. On this day injustice, abuse, and our seemingly unending ambition to destroy ourselves will be confronted and eradicated. But it is also the day of mercy in which God's original intent for this planet as defined in that perfect first week of creation is resurrected. The day of Yahweh is the day when the Creator steps back into our dimension and says, "Enough." It is the day when death dies, the prisoner is freed, the oppressed is delivered, and the oppressor gets his due. This is the *telos* of both the Old and New Testaments.

Predictably, a survey of the biblical text demonstrates that the *yôm Yhwh* is a regular theme in Old Testament prophecy.[14] This day of judgment and mercy is always attended by terrifying signs in the earth and sky; solar eclipses and earthquakes; sounds of thunder and rushing waters; and a huge, heavenly yet earthly army whose task is to bring judgment on all who have colluded against the rule of God. Isaiah 13:2-13 is a classic example:

> Lift up a standard on the bare hill,
> Raise your voice to them,
> Wave the hand that they may enter the doors of the nobles.
> I have commanded my holy ones,
> I have summoned my warriors,
> according to my anger,
> my proud and majestic ones.
> A sound of tumult on the mountains,
> like that of tens of thousands of people!
> The noise of the uproar of kingdoms!
> Of nations gathered together!
> Yahweh of hosts is mustering a host for battle! . . .
> Wail, for the day of Yahweh is near!
> It will come as destruction from the Almighty.
> Therefore all hands will fall limp,

> and every man's heart will melt.
> And they will be terrified....
> Behold, the day of Yahweh is coming,
> cruel, with fury and burning anger,
> to make the land a desolation;
> and he will exterminate the wicked from it!
> The stars of heaven and their constellations
> will not flash their light;
> the sun will be dark when it rises,
> and the moon will not shine its light.
> Thus I will punish the world for its cruelty,
> and the evil ones for their crimes;
> I will put an end to the arrogance of the proud,
> and I will bring down the contempt of the ruthless....
> Therefore I will shake the heavens,
> and make the earth quake from its foundations!
> At the fury of Yahweh of hosts in the day of His burning anger!

In layman's terms, God shows up. In person. To do battle with those who have defiled his inheritance and abused his people. The slaughter will be terrible. The earth and heavens will tremble. But in this great and terrible day the systemic evil that permeates our fallen planet will be purged such that those crushed under the iron fist of injustice will at last know liberation, peace, and prosperity. The Prince of Peace is coming, and his ultimate goal (Greek: *telos*) is the restoration of the perfect world of Genesis 1:

> A civilization without greed, malice, or envy; progress without pollution, expansion without extinction. Can you imagine it? A world in which Adam and Eve's ever-expanding family would be provided the guidance they needed to explore and develop their world such that the success of the strong did not involve the

deprivation of the weak. Here government would be wise and just and kind, resources plentiful, war unnecessary, achievement unlimited, and beauty and balance everywhere.[15]

Thus the "day of Yahweh" may be found in Isaiah 13:6, 9; Jeremiah 46:10; Ezekiel 7:10; 13:5; 30:3; Daniel 2:31-35; Joel 1:15; 2:1, 11; 3:4, 14; Amos 5:18, 20; Obadiah 15; Zephaniah 1:7, 14; and Malachi 4:1, 5, to name but a few of the Old Testament passages.

But the "day of Yahweh" may also be found in the New Testament. Here it is also known as the *parousia* (Greek: "arrival, advent, appearance") or in Christian circles "the second coming" (e.g., Mt 21:33-46 [cf. Is 5]; Mt 24:35-44; Acts 2:20; 1 Cor 5:5; 15:23; 1 Thess 3:13; 5:2; 2 Thess 2:2; Jas 5:8; and 2 Pet 3:10).[16] Why is the "day of Yahweh" in both Testaments? Because the God of the Old Testament *is* the God of the New Testament, and the plan that first set Eden in motion has not changed. The goal has always been God's people living in God's place with full access to his presence. And so in the New Testament we read that God the Son will return as the Captain of Yahweh's hosts and bring with him the deliverance and judgment promised on that glorious but fearful day.

> The sun shall be turned into darkness,
> And the moon into blood,
> before the great and glorious day of the Lord shall come.
> And it shall be, that everyone who calls on the name of the Lord
> shall be saved. (Acts 2:20-21; cf. Joel 2:30-32)

When we understand that the day of Yahweh is the *parousia* of the new covenant and return to 2 Peter and 2 Thessalonians, we return with the lexicon native to the biblical authors. These New Testament writers are speaking in the same idiom as their prophetic forefathers, and they are speaking of an event that any first-century Jew would easily have recognized—the day of Yahweh. Paul makes this explicit in 2 Thessalonians 2:1-3, 7-8:

> Now we request you, brethren, with regard to the coming of our Lord Jesus Christ and our gathering together to Him, that you not be quickly shaken from your composure or be disturbed either by a spirit or a message or a letter as if from us, to the effect that the day of the Lord has come. Let no one in any way deceive you, for it will not come unless the apostasy comes first, and the man of lawlessness is revealed, the son of destruction.... For the mystery of lawlessness is already at work; only he who now restrains will do so until he is taken out of the way. Then that lawless one will be revealed whom the Lord will slay with the breath of His mouth and bring to an end *by the appearance of His coming*. (NASB, emphasis added)

So we see that the imagery of fire and earthquake, the roar of thunder and heavenly disturbances, are common to passages involving the day of Yahweh, and this language is intended to communicate *judgment*, not necessarily annihilation. It is also important to understand that this imagery emerges from a genre known as *apocalyptic* literature—a subcategory of prophetic speech meaning "unveiling" or "revelation" that depicts the end of the world and the inauguration of the kingdom of God in images that are fantastic and sometimes bizarre.[17] This literature is known for its symbolism, mythic imagery, special use of numbers, and periodization of history. Biblical books that are apocalyptic in nature include Daniel in the Old Testament and the New Testament book of Revelation. As Douglas Moo states, "the visions we encounter in these books force us to ask if the prophet is straightforwardly describing the conditions of the new world, or is he using a series of metaphors to describe a state of affairs that have no direct analog to our experience in this world?"[18] Although the continuity between this world and the one to come is not clear to any of us, Moo and a host of New Testament scholars would side with the latter—that these images and metaphors are part of a stock typology for describing the great judgment at the end of the age.[19] In other words, this apocalyptic language does not communicate the complete annihilation of the physical world. Why? Because such a conclusion violates the great arc of redemptive history. Because the imagery is intended to be symbolic. Because the prior judgments rehearsed in the Old Testament do not communicate planetary annihilation, and there are so many other passages in the Old Testament that speak of the restoration and fructification of our fallen planet as a sign of the return of the king. Because even in the great flood of Noah, designed to cleanse the world of evil, which Matthew 24:37 utilizes as a direct analogue for the second coming of the Christ, God preserved the good planet he had made, along with its flora and fauna. And more importantly, because Paul says so.

ROMANS 8:18-25

In the midst of his famous treatment of the inheritance of the saints in Romans 8, we catch a glimpse of Paul's understanding of the fate of our planet. And rather than speaking in terms of obliteration, he speaks in terms of *resurrection*. Paul is probably writing his letter to the Romans from Corinth.[20] He is about to depart for Jerusalem, but he is heavyhearted for the converts in Rome who are struggling with their newfound faith. The questions at hand: Why did we need a new covenant? Who qualifies for membership? Why couldn't the old covenant save us? And most importantly, if the kingdom has come, why are we still poor and persecuted and suffering? And so Paul speaks:

> For I consider that the sufferings of this present time are not worthy to be compared with the glory that is to be revealed to us. For the anxious longing of the creation waits eagerly for the revealing of the sons of God. For the creation was subjected to futility [i.e., frustration], not of its own will, but because of him who subjected it, in hope that the creation itself also will be set free from its slavery to corruption into the freedom of the glory of the children of God.

> For we know that the whole creation groans and suffers the pains of childbirth together until now. And not only this, but also we ourselves, having the first fruits of the Spirit, even we ourselves groan within ourselves, waiting eagerly for our adoption as sons, the redemption of our body. (Rom 8:18-25)

This passage is a poignant presentation of the great arc of redemptive history. I use it often in my teaching to help my students see that the story of redemption doesn't start in Matthew 1:1 or even in Romans 3:23 but starts "in the beginning." Here Eden's role as the blueprint of God's ideal plan, a plan that has been torn asunder by the rebellion of ʾādām, is clear. And the fact that all of redemption's story is leading to the restoration of that perfect plan is clear as well. Throughout the book Paul reflects on the impact of sin on the individual, the community, and the race. In this passage he reflects on its impact on the cosmos. And he reminds his audience that *all creation* has suffered because of humanity's rebellion. As a result, *all* of creation is anxiously awaiting "the revealing of the sons [heirs] of God." Why does creation wait? Because creation itself has been subjected to *frustration*. The Greek in this passage suggests that "creation has been unable to attain the purpose for which it was created."[21] Why? Because the ʾădāmâ (the cultivatable soil) was subjected to ineffectiveness because of the rebellion of ʾādām. God's chosen steward failed in his appointed task, and so the creation over which he had authority was trapped within the self-defeating cycle of humanity's rebellion as well. Creation experiences the same "bondage of decay" as does the human race. And just like the heirs of the kingdom, creation awaits its deliverance.

So how will freedom come to both the cosmos and the children of Adam? Paul abbreviates his answer here but does so with heavily loaded language. Paul elaborates in his discourse in 1 Corinthians 15:42-58. With the return of the last Adam, the children of the first Adam will be born again into that which is imperishable. In Romans this is "the freedom of the glory of the children of God" (Rom 8:21). As Romans 8:23 states, the moment of consummation is our "adoption as sons [heirs]," which is *the redemption of our bodies*. The "glory of the children of God" is that moment when our commitment to Christ in this present age (what New Testament scholars call the "already") is brought to its fulfillment—actualized—by the resurrection of our fallen and broken bodies into a quality of life that is both eternal and able to endure the same dimension as deity (the "not yet"). In other words, the great moment of victory is the moment of resurrection—death is defeated, the curse is repealed, the sons of Adam and the daughters of Eve are finally reconciled with their God and their first home. It is at this moment that the believer's "salvation" is complete (Rom 5:12-21). Now look what Paul does in Romans 8. He juxtaposes the resurrection of *humanity* with the resurrection of *creation*. "The creation itself also will be set free from its slavery to corruption into the freedom of the glory of the children of God" (Rom 8:21). In what many would argue is Paul's theological magnum opus, the apostle argues that the great moment of victory that the believer lives for is the same moment the creation anxiously awaits.[22] When death is defeated and the curse eradicated, the cosmos will also be born again, liberated and healed, freed at last from the chaos of humanity's rebellion.

When we read this passage in class, it is often the first time my students realize that the end goal of the gospel is not simply personal "fire insurance." They are usually a bit stunned (and always totally jazzed) to find out that their personal story of salvation is actually only one small part of a panoramic master plan to restore *all* of creation through the work of the Christ. In the words of Douglas Moo,

> If creation has suffered the consequences of human sin, it will also enjoy the fruits of human deliverance. When believers are glorified, creation's "bondage to decay" will be ended, and it will participate in the "freedom that belongs to the glory" for which Christians are

destined. Nature, Paul affirms, has a future within the plan of God. It is destined not simply for destruction but for transformation.[23]

REVELATION 21:1 AND REVELATION 22:1-2

The book of Revelation offers an additional glimpse of the master plan. Here what Christians name "heaven" is identified as "a new heaven and a new earth," a "new Jerusalem" coming down out of heaven from God, "made ready as a bride adorned for her husband" (Rev 21:1-2 NASB). John describes heaven with all of the past beauties of Jerusalem, purified and amplified until we are gazing at a city that sparkles like jasper "as clear as crystal" (Rev 21:11), with gates made of pearls, guarded by angels, perfectly square, as was the holy of holies, and made of pure gold (Rev 21:18). The gates never close, because there is no danger there (Rev 21:25). There is no need of sun or moon or lamps, because "the glory of God illuminates the city, and the Lamb is its light" (Rev 21:23). In this place, the cosmic river of Eden is free to flow from the throne of the rightful king (Rev 22:1; cf. Ezek 47:1-12), and Eden's tree of life has multiplied such that it lines the central street of the city (Rev 22:1-2).

> There will no longer be any curse; and the throne of God and of the Lamb will be in it, and His bond-servants will serve Him; and they will see His face, and His name will be on their foreheads. And there will no longer be any night; and they will not have need of the light of a lamp nor the light of the sun, because the Lord God will illumine them; and they will reign forever and ever. (Rev 22:3-5 NASB)

As I discuss at length in *The Epic of Eden*, the iconography John deploys makes it clear that unlike the disembodied existence most Christians envision "heaven" to be (an existence in which we are destined to float around the heavenlies for all eternity playing harps among the clouds), "heaven" is in reality Eden restored.[24] By describing heaven with Eden's sacred river and tree of life, the New Testament writers are intentionally forging connections for their readers. They are leaving us "theological breadcrumbs" to lead our minds back to Eden. The books of Romans and Revelation are telling us that just as our bodies will be raised as living flesh and blood, our heaven will need to accommodate such a corporeal identity. Thus, although missed by too many readers, the New Testament is teaching us that "heaven" is this very earth resurrected, healed of its scars, and washed clean of its diseases. As Gregory Beale states, it is "an identifiable counterpart to the old cosmos and a renewal of it, just as the body will be raised without losing its former identity."[25] And although it is true that the continuity between this world and the next is difficult to define, the fact that Paul dares to associate the final destiny of this planet with the ultimate expression of a believer's identity as the redeemed heir of God (i.e., the resurrection of the body) speaks volumes regarding the intrinsic value that God places on this planet.

All said, although it is true that the audience of the New Testament is more urban than that of the Old Testament, and as a result we hear far less about agriculture and pastoralism than we do in the Old. And although the theocracy of the nation of Israel is no longer functioning in first-century, Roman-occupied Judea, and therefore federal law is no longer God's law. And although the corpus of the New Testament has a nearly singular focus to make plain the character of the new ʾādām, and therefore offers us less in the way of land tenure and creature care, this inquiry makes it clear that the New Testament continues to embrace and reiterate the message of the Old: "For by Him all things were created, both in the heavens and on earth, visible and invisible, whether thrones or dominions or rulers or authorities—all things have been created by Him and for Him" (Col 1:16). In other words, in the New Testament the garden (and the widow and the orphan and the creature) still belong to God. God still intends that the resources of this planet be utilized for his purposes. And according to the apostle Paul and John the Revelator, God's most central purpose for his garden is to redeem it.

DISCUSSION QUESTIONS

1. Would you concur that the church's only mission is the conversion of souls?

2. Do you believe the church is also responsible for charity and service toward the widow and the orphan (even the unsaved widow and orphan)?

3. Do you think the New Testament would agree with the Old Testament as regards sustainable use of the land and humane treatment of livestock?

4. Where do you think the assumption that it is ethically appropriate to use the earth's resources as aggressively as possible to accomplish what "really matters" (the conversion of souls) has come from?

5. Having read this chapter, would you identify environmental stewardship as peripheral or alien to the theological concerns of the Bible?

CONCLUSION

How Should We Then Live?

> *I used to think that the top environmental problems*
> *were biodiversity loss, ecosystem collapse and climate change.*
> *I thought that thirty years of good science could address these*
> *problems. I was wrong. The top environmental problems are*
> *selfishness, greed and apathy, and to deal with these*
> *we need a cultural and spiritual transformation.*
> *And we scientists don't know how to do that.*
>
> GUS SPETH, CHAIRMAN OF THE COUNCIL ON
> ENVIRONMENTAL QUALITY UNDER
> PRESIDENT JIMMY CARTER

IN THIS BRIEF STUDY, I have attempted a biblical theology of environmental stewardship. My method has been consistent with the task of biblical theology throughout the ages, to submit the topic to a survey of the biblical text. I have asked the questions: Do I see this particular precept *systematically* represented in the text as an aspect of the character of God? Or is this value limited to a marginal representation in the text via the particularities of situational ethics? We have seen that Scripture speaks to this topic repeatedly and systematically, and contrary to what we may have thought when we first opened this little book, the stewardship of this planet is *not* alien or peripheral to the message of the gospel. Rather, our rule of faith and praxis has a great deal to say about the subject. We have learned that God owns this planet and that both the garden and Israel were "land grants" offered to humanity for their responsible stewardship. We have seen that the themes of sustainable land use, humane treatment of livestock, care for the wild creature, respect for the flora and fauna of our leased land, and care for the widow and orphan are reiterated from Eden to the new Jerusalem. So from the perspective of Scripture—yes! the Bible does speak to these concerns.

So now for the most challenging question: Where should we as Christians position ourselves with regard to these truths? As discussed in the introduction, the message of environmental concern is politically charged for many. But when we put aside national politics and focus in on kingdom politics, a new picture emerges. Stewardship of this planet is not a Republican versus Democrat conversation. This is not the NRA versus Planned Parenthood or "liberal" versus "conservative." Rather, what we are discussing here is the call to be a Christ-follower in a fallen world. Of all the voices and all the "facts" that are calling for our allegiance in the many arenas of environmental thought, for the citizen of the kingdom of God, the voice of Scripture must surpass them all.

My study of the nation of Israel made it clear that the first kingdom of God required its citizens to serve and protect the garden. In Israel neither economic expansion, national security, nor even personal economic viability, were legitimate justification for the abuse of the land, the poor, or the domestic or wild creature. Rather, all of Israel's laws of land, tree, and creature communicate the same premise as that in Eden: Israel was a tenant on God's good land, a steward. The land, its produce, and its inhabitants belonged to God, not humanity. And each member of Israel's society stood responsible before God for their care of his resources. We saw that the broader testimony of the Old Testament reinforces this same message. God takes pleasure in his creation. He has designed it, provided for it, and his expectation is that his people will respect and protect it. If I were to summarize the message of the Old Testament regarding creation care into a single proverb it would be this:

> *The earth is the Lord's and all it contains;*
> *you may make use of it in your need,*
> *but you shall not abuse it in your greed.*

We also found that Israel's attitude toward the enduring fertility of their land, its wild residents, and the well-being of their livestock stood in significant contrast to the practices of other societies of their time. Egypt, Mesopotamia, and the Aramaeans were well known for their environmental terrorism in warfare. Assyrian iconography celebrated the wanton slaughter of the wild creature, and it is broadly believed that a contributing factor to the collapse of the Mesopotamian civilization was agricultural sterility resulting from the failure to fallow. According to my training as an academic, I should be looking for an explanation of Israel's distinctive mindset in the sociological realties of the evolution of its culture. Perhaps Deuteronomy's concern for the long-term environmental impact of their civilization on the land is the result of their uniquely challenging geographical setting, the psychological impact of their dependence on dry farming as opposed to irrigation-based agriculture as in Egypt and Mesopotamia? Perhaps Israel's unique perspective grew out of their egalitarian societal structure, unlike the empires surrounding them? Or perhaps Israel's law is a standard ethnocentric reaction against the practices of "the other"? My study of this topic has convinced me that the answer to this question is not to be found in Israel's social formation or egalitarian politics. Rather, it seems to me that Israel's distinctive perspective is instead a reflection of the character of their God. A reflection that critiqued and censured their culture and their economy just as much as it does ours. Living sustainable lives that embrace restraint and charity and the rhythm of Sabbath, lives that honor the long-term fertility of the land and well-being of both the domestic and the wild creature, was no easier for Israel than it is for us. Just like us, Israel struggled with the competing demands of a diverse society, insufficient yields, property loss, land tenure, poverty, and taxes. But underlying their response to these issues was one central tenet: this land, these creatures, are not ours. They are on loan to us. We must manage them well so that each is preserved. And we must take God at his word that in response to our obedience, he himself will bring about the increase (Deut 30:9). Short-term, desperation management that exhausts current resources in answer to the cry of the urgent was not acceptable in Israel, and it cannot be acceptable to us either.

Thus, of all the messages regarding creation care that might be attributed to Scripture, one seems incontrovertible to me: the garden and its creatures are not ours, they are his. Our God-ordained task, before the fall, was to care for his garden, to serve it (*lĕʿobdāh*) and to guard it (*lĕšomrāh*; Gen 2:15). Our fallen race has instead chosen to use its superior gifts to exploit and abuse. In our greed we have taken what we wanted with no concern (often no thought) as to what the consequences of our behavior might be. At this point, the statistics quantifying those consequences are truly staggering: countless waterways poisoned, tens of thousands of species lost, millions of acres decimated, unfathomable quantities of trash. Humanity was created and commanded to serve and to protect, yet humanity has instead ravaged the garden. And like the results of *ʾādām*'s choice in the arena of human relationships, in the arena of our relationship with creation, the results are all around us.

But God's people are called to be different. We live for the day when the creation itself will be set free from "its slavery to corruption into the freedom of the glory of the children of God" (Rom 8:21 NASB). Our God-given calling is to serve as witnesses to the fact that "all things have been created by Him and for Him" (Col 1:16). Thus, in this fallen world, the role of the redeemed community is to live our lives as an expression of another kingdom, to reorient our values to those of our heavenly Father, to live our lives as Adam and Eve should have, as Jesus Christ has. Our calling is to demonstrate with our *lives* "what the will of God is, that which is good and acceptable and perfect" (Rom 12:2 NASB). How can we avoid this message, then, that it is our responsibility as redeemed humanity to live in such a way that the intentional stewardship of God's creation is evident in our lives?

And so I return to my proverb:

> *The earth is the Lord's and all it contains;*
> *you may make use of it in your need,*
> *but you shall not abuse it in your greed.*

I firmly believe that our current environmental crisis is not the result of need; it is the result of greed. Returning to Gus Speth's quotation that opened this chapter, I want us to pay attention to the fact that this long-time veteran of environmental activism, cofounder of the National Resource Defense Council, founder of the World Resources Institute, CEO of the UN Development Programme, and President Carter's chair of the Council on Environmental Quality, has come to the same conclusion. I cannot imagine a person who has a better "inside scoop" on environmental politics and policy than Speth. But after thirty-five years of working in the highest echelons of environmental activism and national policy, Speth is convinced that the real issue behind environmental degradation is not science or politics—it is morality. A morality that allows government to promote "money power over people power."[1] A morality that allows unrestrained profit to prevail over all other societal goods. A morality that allows the inherent balance of a democratic system to be held hostage by "big money." Speth's critique of government might not surprise you. But what should surprise you is that this man, who has attained more influence in his career than most of us could ever dream of, *is asking for help*. Why? Because he has realized that scientists and activists cannot change the morality of a nation. That is where we come in. Because the community of the redeemed *can*.

At its very best, the church has led the way on abolition, temperance, homelessness, orphan and foster care, medical services for the least of these, and civil rights. At our best we have built more orphanages and hospitals than any other single organization on this planet. At our best we have taken our God-ordained self-sacrificial posture to places such as Madagascar and Haiti and served as educators, botanists, and midwives. We have embraced our role as the moral compass of society, confronted corruption, and defended the voiceless. Can we do it again? Can we step up and lead on the topic of environmental stewardship? Douglas Moo says it this way: "The 'not yet' of a restored creation demands an 'already' ethical commitment to that creation now among God's people."[2] What he means is that our identity as a witness to God's character in this fallen world demands that we live our lives as animated representations of what God's kingdom *will* be. As I told a sobered and impassioned gathering of young PhDs at the Evangelical Theological Society in 2012, I do not anticipate that the church will be able to *fix* all (or even most) of the environmental woes of our planet any more than we will be able to end every war, adopt every orphan, or free every young woman trapped in the sex trade. But I do believe that we can stand boldly with the opposition. Yes, when we speak of living lives of restraint and stewardship we are articulating a value system that is foreign to our fallen world—as is expressly obvious by the relational and environmental mess that surrounds us. But the fact that our message is countercultural in no way releases us from the prophetic task at hand. A light in the darkness, leaven in the lump: that is who we are. An audacious (but prophetic) Margaret Mead is often credited with this quotation: "Never doubt that a small group of thoughtful, committed citizens can change the world; indeed, it's the only thing that ever has."

In sum, I am completely convinced that the redemption of all creation *is* the gospel. Therefore, creation care is not merely a message of social justice, a wise approach to life on this planet, or a political action item. It is instead a life posture that reflects the character of God and embodies the *telos* of his plan. Like all of the fallout of Eden, the only true solution to our dilemma is the gospel—the message of transformed lives, living in alliance with God's strategic plan. The apostle Paul says it this way, that our calling is to demonstrate with our *lives* "what the will of God is" (Rom 12:2). What is the will of God regarding creation? "Then Yahweh Elohim took the human and put him into the garden of Eden to tend it [lĕʿobdāh] and protect it [lĕšomrāh]" (Gen 2:15).

The introduction of this book asked the question: Can a Christian be an environmentalist? My answer is, how could a son of Adam or a daughter of Eve, redeemed and transformed by the second Adam to live eternally in the resurrected Eden, be anything else?

Appendix

RESOURCES FOR THE RESPONSIVE CHRISTIAN

In an article published in *Sierra* in April 2018, climate-science communicator Eric Holthaus confessed that his research into climate change had landed him in a therapist's chair in need of help with unmanageable anxiety. "Like many people who care about the fate of the planet, I've spent most of the past year alternating between soul-crushing despair and headstrong hope."[1] According to Holthaus, the therapist was a bit taken aback by the intensity of this scientist's emotional dread. His advice? "Do what you can." Holthaus reports that this very simple advice really helped him "to realize something important: We are all in this together."

At a Seminary Stewardship Alliance meeting in 2014, I was asked to offer a devotional on Romans 8. Standing in front of a deeply committed community of professors, activists, and lay leaders, like Hothaus I was struck by how deeply discouraging the battle can be. Whole species are slipping through our hands as I write, mountains are being blown into oblivion as you read these words, irreplaceable wilderness is groaning under the machinery of shortsighted industrial development. Thus, as I concluded my devotional on Paul's powerful words about the resurrection of this planet, I told my audience what I am telling you. If Paul were among us now, I think he would remind us that we must be encouraged.

Even though our efforts, like the cosmos, seem to be subjected to frustration, know that they have been subjected in *hope*. The hope of a plan that cannot fail. The hope of a God who stands behind and before his people. A God who stands as our advocate, advancing and empowering *our* efforts on his behalf. So as you take up this task, know that as with everything else in the Christian life, we hope for what we cannot see, and with perseverance we eagerly wait (and work!) for it. As my friend and past colleague ethicist Christine Pohl was often heard to say: "Small moves against the darkness." That is your task. And every small move ... matters.

ENVIRONMENTAL STEWARDSHIP: HOW TO TAKE ACTION

As you look through this list, choose *one* thing you can do this week and start. Consider adding a second next month, and maybe another three months from now. Choose the easy things first, and get rolling!

1. Get informed! Join a responsible environmental organization. This way *you* commit some portion of your resources to the cause and *they* keep you informed. Among my favorites are the **Sierra Club** and the **Nature Conservancy**. The Humane Society and the Defenders of Wildlife are organizations that specialize in creature care. Make sure you check out the local chapter in your town as well.

2. Vote your *informed* conscience! A nice feature of the *Sierra Club* magazine is that it always has a section on local government representatives come voting season. I of course never agree with all of their recommendations, but this resource helps me identify which of my representatives have been voting for what. It also helps me contact them and speak up.

3. Begin to vote with your finances! What you choose to buy does more to change the face of industry than any other single action.

This includes what wood your new dining room table is made of, where your meat came from, and your investment portfolio.

- a. Buy organic and recycled as often as you can. By this you communicate to the agricultural industry that you as a consumer *want* responsibly grown food products. Look for "free range," "grass fed," and "humanely raised" labels on your meat products.
- b. Buy "free range" (not just "cage free") eggs. By this you communicate that you care how the hens that produce our eggs live out their lives. See the Humane Society's website to get more information: humanesociety.org/issues/confinement_farm/facts/battery_cages.html.
- c. Buy "reduced packaging." Less packaging means less processing, less waste, less trash, less land and resources committed to processing trash.

4. Live with restraint. Every time you buy a smaller house, a smaller car, less recreational equipment, and less junk you preserve resources (not the least being your credit rating!).

5. Learn how and where to **recycle responsibly**. The information is easily available online. In 2013, Americans generated about 254 million tons of trash and recycled and composted about 87 million tons of this material, equivalent to a 34.3 percent recycling rate (municipal solid waste, US EPA archives).

6. As approximately 18 percent of the trash your household produces is **compost**, think seriously about designating some portion of your yard for yard clippings and fruit and vegetable waste. Our family composts everything but meat.

7. Make sure your school, church, and office are recycling—if they aren't, help them. Shred-it remains a leading document-shredding and paper-recycling company that is worth a look (see shredit.com).

8. Look online to see whether you have a local wildlife rehabilitator. Get their number in your phone. Maybe make a donation. I just carried an oil-soaked loon home from a morning walk, and because of our local wildlife group, he's doing just fine.

9. Address your own consumption of energy in your room, your home, your office. Little changes make a big difference. You might even schedule an energy audit with your local utility. Simple fixes such as closing curtains in summer, caulking around windows and doors, and switching to LED light bulbs save a lot of energy. A three-degree adjustment to your thermostat reduces heating and cooling consumption (and bills!) by 10 percent.

   a. Did you know you can *rent* solar panels?

   b. Do you remember how nice a clothesline can be?

10. Attend to your automobile! In your family's life, more than likely your biggest use of fossil fuel is your car. The size of your vehicle, the number of vehicles, the number of miles you drive, updated maintenance on those vehicles, are all enormously significant. Buy smaller, fewer, and with better gas mileage. Your grandchildren and your savings account will thank you.

11. Join a community supported agriculture group (CSA). By buying your produce (and meat) from a CSA, you offer the family farm a niche in our economy, support environmentally responsible agriculture, and boost your family's health. Just punch "CSA" or "Community Supported Agriculture Group" into Google to find a CSA in your area to order fresh, local produce online.

12. Give up your chemical lawn service and be very restrained in the use of pesticides. The link between these chemicals in our food

chain and human cancer is clear. Consider an outfit such as Gardens Alive (gardensalive.com) for making your lawn beautiful and keeping ants out of your sugar canister.

13. Plant **native trees and plants** in your yard. Indigenous plant species grow better and faster, need less water, are resistant to local pests and diseases, and attract wildlife.

14. Think about **water** consumption in your landscape.

    a. Soaker hoses on a timer used at night are the best way to water everything.

    b. Sprinklers during daylight hours are an invitation to evaporation and are not good for your checkbook or your plants.

    c. Think about rain barrels at the base of your drain spouts.

15. Read Nancy Sleeth's very user-friendly *Go Green, Save Green: A Simple Guide to Saving Time, Money, and God's Green Earth*. It is available as an ebook on Amazon. The first two chapters and the "Church" chapter are available under "church resources" on the Blessed Earth website, blessedearth.org.

16. Support **environmental missionary** efforts that seek to give witness to Christ while restoring indigenous habitat. Groups such as Plant with Purpose (plantwithpurpose.org) and Red Island Restoration (redislandrestoration.com) are worthy of your attention.

17. Are you a pastor or do you want to start a "**green team**" at your church?

    a. Keep in mind that a church is an enormous public facility that uses all sorts of resources. But rarely does a building committee or pastor have training in sustainable-energy use, native planting, water conservation, recycling, and so on. These questions need to be addressed as a community of faith builds its worship center. Be the person who offers to help.

b. Blessed Earth offers a twelve-part DVD Bible study, *Serving God, Saving the Planet*. Matthew and Nancy Sleeth report that after a dozen years in the fight, it's the most successful way they've seen to start and maintain a church group committed to making practical changes at the personal, church, and community levels (blessedearth.org).

NOTES

1 CREATION AS GOD'S BLUEPRINT

[1] For a fuller treatment of this reading of Gen 1, see Sandra L. Richter, *The Epic of Eden: A Christian Entry into the Old Testament* (Downers Grove, IL: IVP Academic, 2008), 92-118; Henri Blocher, *In the Beginning: The Opening Chapters of Genesis* (Downers Grove, IL: InterVarsity Press, 1984), 15-38; Meredith G. Kline, *Kingdom Prologue: Genesis Foundations for a Covenantal Worldview* (Eugene, OR: Wipf & Stock, 2006), 38-43.

[2] Ludwig Koehler, Walter Baumgartner, M. E. J. Richardson, and J. J. Staum, *Hebrew and Aramaic Lexicon of the Old Testament* (Leiden: Brill 2000), s.v. "רדה" (1190). Although most often translated "to rule," the argument has been made that "the basic meaning of the verb ... actually denotes the travelling around of a shepherd with his flock" (*Hebrew and Aramaic Lexicon of the Old Testament*, s.v. "רדה" [1190]). This latter translation obviously shares the connotations of authoritative leadership, but also communicates the image of a steward who is on the move throughout his territory making sure all is well with his charges. This translation conforms nicely with the commission of the first humans. For other occurrences of *rādâ* see Lev 25:43, 46, 53; Num 24:19; and 1 Kings 5:4.

[3] For a detailed treatment of the ancient Near Eastern contexts of this discussion and the biblical theological implications, see Catherine McDowell, *The Image of God in the Garden of Eden: The Creation of Humankind in Genesis 2:5–3:24 in Light of the* mīs pî, pīt pî, *and* wpt-r *Rituals of Mesopotamia and Ancient Egypt*, Siphrut 15 (Winona Lake, IN: Eisenbrauns, 2015). See also Sandra L. Richter, *The Epic of Eden: Isaiah,* A One Book Curriculum (Franklin, TN: Seedbed, 2016).

[4] *Hebrew and Aramaic Lexicon of the Old Testament*, s.v. "כבש" (460). "Subjugate" is a typical translation of this verb. It is found throughout the Bible in contexts in which a new people group or conquering army has moved into new territory and taken possession of it (e.g., Num 32:22, 29; Josh 18:1; 2 Sam 8:11). For example, David asks his cabinet, Has God "not given you rest on every side? For He has given the inhabitants of the land into my hand, and *the land is subdued* [*kābaš*] before the LORD and before His people" (1 Chron 22:18 NASB). Note that this verb is typically used regarding land, not animate creatures.

[5] As Meredith G. Kline states: "A hierarchical pattern of dominion can be traced through the creational record, a pattern of ascending consecration with the Sabbath as its capstone.... Within the first three day-frames is described the origin of three vast spheres over which rule is to be exercised. Then in day-frames four through six the rulers of each of these spheres is presented in proper turn, each arising at the divine behest and ruling by divine appointment. But the rising chain of command does not stop with the six days; it ascends to the seventh day, to the supreme dominion of him who is Lord of the Sabbath" (Kline, *Kingdom Prologue*, 38).

[6] *Hebrew and Aramaic Lexicon of the Old Testament*, s.v. "משל" (647-48). As in Dan 11:39, when combined with the preposition בְּ this verb communicates placing someone into an office of authority.

[7] See Richter, *Epic of Eden*, 69-91.

⁸For additional occurrences of the verb *kābaš* utilized with a land grant, see n. 3 as well as Josh 18:1 and 2 Sam 8:11.

⁹*Hebrew and Aramaic Lexicon of the Old Testament*, s.v. "עבד" (773). The most essential meaning of this verb is "to serve; to work." When in the context of cultivatable land, it is typically translated "to till"; when with an animal, "to work with."

¹⁰*Hebrew and Aramaic Lexicon of the Old Testament*, s.v. "שמר" (1581-84). A very common verb utilized in an array of contexts from Northwest Semitic through the Dead Sea Scrolls into Middle Hebrew. The essential meaning is to "keep watch over"; "take care of"; "keep and/or do something carefully." This is the verb from which "watchman" derives.

¹¹*Hebrew and Aramaic Lexicon of the Old Testament*, s.v. "נוח" (679). The second meaning of the *hiphil* of this form is utilized for installing animated statues of the gods in their temples. See "The Servant and the Idol," in my Bible study curriculum, *Epic of Eden: Isaiah*.

¹²See Douglas J. Moo and Jonathan A. Moo, *Creation Care: A Biblical Theology of the Natural World* (Grand Rapids: Zondervan, 2018), 147-52, for further discussion.

¹³Douglas J. Moo, "Nature in the New Creation: New Testament Eschatology and the Environment," *Journal of the Evangelical Theological Society* 49 (2006): 461.

¹⁴Richter, *Epic of Eden*, 102-12.

¹⁵Richter, *Epic of Eden*, 114-16.

¹⁶The first civilian hospital was built by Christians in the Roman Empire (Virginia Smith, *Clean: A History of Personal Hygiene and Purity* [New York: Oxford University Press, 2007], 142).

¹⁷Timothy Lawrence Smith, *Revivalism and Social Reform: American Protestantism on the Eve of the Civil War* (New York: Harper & Row, 1965), 173-74. Smith's work is a classic resource on the relationship between revival and social reform in the United States. See in particular his chapters "The Churches Help the Poor," 163-77, and "The Spiritual Warfare Against Slavery," 204-24.

¹⁸See the Union Rescue Mission website at https://urm.org/solution/.

2 THE PEOPLE OF THE OLD COVENANT AND THEIR LANDLORD

¹For a discussion of Deuteronomy as Israel's constitution (*politeia*), see S. Dean McBride, "Polity of the Covenant People: The Book of Deuteronomy," in *A Song of Power and the Power of Song: Essays on the Book of Deuteronomy*, ed. Duane L. Christensen (Winona Lake, IN: Eisenbrauns, 1993), 62-77.

²See Sandra L. Richter, *The Epic of Eden: A Christian Entry into the Old Testament* (Downers Grove, IL: IVP Academic, 2008), 69-91; Moshe Weinfeld, "בְּרִית *bᵉrît*," in *Theological Dictionary of the Old Testament*, ed. G. Johannes Botterweck and Helmer Ringgren, trans. John T. Willis et al. (Grand Rapids: Eerdmans, 1975), 2:267; and Bill T. Arnold and Bryan E. Beyer, eds., *Readings from the Ancient Near East: Primary Sources for Old Testament Study* (Grand Rapids: Baker Academic, 2002), 96-103.

³"To place one's name" is an ancient idiom communicating that a king had inscribed his name and/or inscription on a monument or building. "Placing one's name" served to designate the building or monument, or the territory it marked, as belonging to the king. See Sandra L. Richter, *The Deuteronomistic History and the Name Theology: lᵉšakkēn šᵉmô šām in the Bible and the Ancient Near East*, Beihefte zur Zeitschrift für

die alttestamentliche Wissenschaft 318 (Berlin: de Gruyter, 2002); and Richter, "The Place of the Name in Deuteronomy," *Vetus Testamentum* 57 (2007): 342-45.

⁴The law of Ex 22:28 has changed here. Rather than the firstborn being sacrificed on the eighth day, now—with the dispersal of the Israelites throughout Canaan—the sacrifice of the firstlings occurs only once yearly.

⁵In Israelite society the economic function of ovines greatly outnumbered bovines, with ovines serving as the primary source of milk and meat and bovines associated chiefly with the cultivation of cereals (Baruch Rosen, "Subsistence Economy in Iron Age I," in *From Nomadism to Monarchy: Archaeological and Historical Aspects of Early Israel*, ed. Israel Finkelstein and Nadav Na'aman [Jerusalem: Israel Exploration Society, 1994], 339-49; cf. Prov 14:4. See also Rosen, "Subsistence Economy of Stratum II," in *'Izbet Ṣarṭah: An Early Iron Age Site near Rosh Ha'ayin, Israel*, ed. Israel Finkelstein, BAR International Series 299 [Oxford: B.A.R., 1986], 180).

⁶Patrick D. Miller, *The Religion of Ancient Israel* (Louisville, KY: Westminster John Knox, 2000), 120-21.

⁷Miller, *Religion of Ancient Israel*, 119.

⁸Ryan Strebeck of the Strebeck Family Ranch (personal communication, September 8, 2008).

⁹Ann Bell Stone of Elmwood Stock Farm (personal communication, September 11, 2008); cf. Sandra L. Richter, "Environmental Law in Deuteronomy: One Lens on a Biblical Theology of Creation Care," *Bulletin for Biblical Research* 20 (2010): 355-76.

¹⁰Rosen, "Subsistence Economy of Stratum II," 177-78. See Timothy S. Laniak, *Shepherds After My Own Heart: Pastoral Traditions and Leadership in the Bible*, New Studies in Biblical Theology 20 (Downers Grove, IL: InterVarsity Press, 2006), 42-57; Gudrun Dahl and Anders Hjort, *Having Herds: Pastoral Herd Growth and Household Economy*, Stockholm Studies in Social Anthropology 2 (Stockholm: Dept. of Social Anthropology, University of Stockholm, 1976); Brian Hesse, "Animal Husbandry and Human Diet in the Ancient Near East," in *Civilizations of the Ancient Near East*, ed. Jack M. Sasson (1995; repr. in 2 vols., Peabody, MA: Hendrickson, 2006), 1:203-22. See Sandra L. Richter, "The Question of Provenance and the Economics of Deuteronomy," *Journal for the Study of the Old Testament* 42 (2017): 23-50; Sandra L. Richter, "The Archaeology of Mt. Ebal and Mt. Gerizim and Why It Matters," in *Sepher Torath Mosheh: Studies in the Composition and Interpretation of Deuteronomy*, ed. Daniel I. Block and Richard L. Schultz (Peabody, MA: Hendrickson, 2017), 311-16.

¹¹Although the book of Deuteronomy does not provide a detailed list of each tribe's inheritance, the frequent references to the Levites and their lack of inheritance (e.g., Deut 10:9; 12:12; 14:27, 29; 18:1; 29:8), the mention of cities of refuge (Deut 4:41; 19:2, 7), the discussion of the inheritance of the Transjordanian tribes (Deut 29:8), and the periodic topographical references in the tribal blessings of Deut 33 make it clear that the writer is presupposing a distribution of the land into tribal territories as is detailed throughout Josh 12–24. In light of Yahweh's ownership of this land, the patrimony of the Israelite could not be sold in perpetuity (Lev 25:13-17, 23; cf. Is 5:8). Thus the citizenry was not allowed to abuse each other or the land by means of the self-serving acquisition and sale of real estate. See Oded Borowski on private land tenure in his *Agriculture in Iron Age Israel* (Boston: American Schools of Oriental Research, 2002), 23-26.

[12] As J. Gordon McConville states, Deut 5:12 is a "conscious re-presentation" of Ex 20:8, which mandates rest for the land as well as its creatures (*Deuteronomy*, Apollos Old Testament Commentaries 5 [Leicester, UK: Inter-Varsity Press, 2002], 121-22, 128).

[13] David C. Hopkins, "Life on the Land: The Subsistence Struggles of Early Israel," *Biblical Archaeologist* 50 (1987): 185. Most organic farmers still practice this sort of fallowing for the same reasons.

[14] See "Mixed Crop-Livestock Farming: A Review of Traditional Technologies Based on Literature and Field Experience," Food and Agriculture Organization of the United Nations, FAO Animal Production and Health Papers 152, www.fao.org/3/Y0501E/y0501e00.htm#toc.

[15] For the role of livestock manure in this system, see Rosen, "Subsistence Economy in Iron Age I," 344-45, as well as Borowski, *Agriculture in Iron Age Israel*, 145-48.

[16] Borowski, *Agriculture in Iron Age Israel*, 95. According to Professor Patricia Muir of the Department of Botany and Plant Pathology at Oregon State University, "Conversion to cropland is almost universally associated with a rapid decrease in soil organic matter and soil nitrogen content." She notes that "a dramatic example of this is in the midwestern US, whose prairie soils have lost 1/3-1/2 of their organic material since they began being cultivated" (Patricia S. Muir, "Consequences for Organic Matter in Soils," http://people.oregonstate.edu/~muirp/orgmater.htm).

[17] For the history of Mesopotamia's failure to attend to these principles, see Thorkild Jacobsen and Robert M. Adams, "Salt and Silt in Ancient Mesopotamian Agriculture," *Science* 128 (1958): 1251-58; cf. Norman Yoffee, "The Collapse of Ancient Mesopotamian States and Civilization," in *The Collapse of Ancient States and Civilizations*, ed. Norman Yoffee and George L. Cowgill (Tucson: University of Arizona Press, 1991), 53; cf. Borowski, *Agriculture in Iron Age Israel*, 148. Jacobsen and Adams point out that "at about 2400 B.C. in Girsu a number of field records give an average yield of 2537 liters per hectare—highly respectable even by modern United States and Canadian standards. This figure had declined to 1460 liters per hectare by 2100 B.C., and by about 1700 B.C. the recorded yield at nearby Larsa had shrunk to an average of only 897 liters per hectare" (Jacobsen and Adams, "Salt and Silt," 1252). Yoffee comments that "there may well have been a decision to abandon or shorten the period of fallow on lands the Crown controlled, thereby providing short-term fiscal relief, since the lands would initially provide more grain, but ultimately there would result a loss in productivity" ("Collapse of Ancient Mesopotamian States," 53). See Marvin A. Powell, "Salt, Seed, and Yields in Sumerian Agriculture: A Critique of the Theory of Progressive Salinization," *Zeitschrift für Assyriologie und Vorderasiatische Archäologie* 75 (1985): 7-38. Yoffee theorizes that the ever-present threat of salination of the Mesopotamian soils due to extensive irrigation was exacerbated by Hammurabi's centralization of the realm.

[18] See Lawson G. Stone, "Worship as Cherishing YHWH's World in Leviticus" (paper presented at the Annual Meeting of the Institute for Biblical Research, New Orleans, November 21, 2009).

[19] The literature related to this topic is immense and multidisciplinary. For an introduction, see Scott Sabin, "Environmental Emigration: The World on Our Doorstep," *Creation Care* 37 (Fall 2008): 37-38 (Sabin is the executive director of Plant with Purpose, an organization dedicated to restoring indigenous habitats in order to restore

the economic and social well-being of the marginalized. https://plantwithpurpose.org/board/). See also Pavan Sukhdev, "The Economics of Ecosystems and Biodiversity—TEEB," interim report, May 29, 2008, with overview at EurekAlert!, May 29, 2008, www.eurekalert.org/pub_releases/2008-05/haog-teo052908.php. The website for the global mission of TEEB is http://ec.europa.eu/environment/nature/biodiversity/economics/.

[20] Lauren F. Winner, *The Mudhouse Sabbath* (Brewster, MA: Paraclete Press, 2003), 10.

[21] Borowski, *Agriculture in Iron Age Israel*, 148.

[22] Borowski believes that he has found evidence for crop rotation in ancient Israel in the taboo of *kil'ayim*, "mixture of two kinds," recorded in Deut 22:9 and Lev 19:19 (*Agriculture in Iron Age Israel*, 150-51). See Gideon Ladizinsky, "Origin and Domestication of the Southwest Asian Grain Legumes," in *Foraging and Farming: The Evolution of Plant Exploitation*, ed. David R. Harris and Gordon C. Hillman (London: Unwin Hyman, 1989), 374-89; cf. Borowski, *Agriculture in Iron Age Israel*, 95. Note that unlike Israelite law, federal law in the United States has actually discouraged crop rotation and fallow cycles. Patricia Muir states, "For many years, US government farm policies essentially blocked farmers from rotating. To receive full crop subsidies and other financial supports, growers had to commit acreages to certain crops, which made rotation a financial liability.... The 1990 and 1996 Farm Bills eased these restrictions a bit, but there are still restrictions" ("Consequences for Organic Matter in Soils"). Muir argues that such growing practices essentially *necessitate* the use of chemical fertilization. Although chemical fertilizers do provide standardized results and immediate resuscitation of the soil, in the long run, they strip the soil of its fecundity. See Muir's summary of the impact of organic versus nonorganic fertilizers on long-term soil structure and fertility ("Consequences for Organic Matter in Soils").

[23] As Muir points out, organic material, i.e., humus, in soil is necessary for water-holding capacity, aeration, maintenance of beneficial soil organisms, and the input of natural fertilizers ("Consequences for Organic Matter in Soils").

[24] Hopkins, "Life on the Land," 185.

[25] In Lev 26:34-35, 43, Yahweh states that, because Israel did not keep the law of the Sabbath fallow, he will give those Sabbaths back to the land over the course of the exile.

[26] Note that the law codes of both Egypt and Mesopotamia place great stress on the duty of the tenant to keep the soil in good working order (Christopher Eyre, "The Agricultural Cycle, Farming and Water Management in the Ancient Near East," in Sasson, *Civilizations of the Ancient Near East*, 1:185).

[27] William Gaud was the administrator of the Department of State's Agency for International Development, and the address in question was made on March 8, 1968, to the Society for International Development. Due to their essential role in the food systems of most Third World nations, wheat and rice were the first two grains targeted by this. Soon after high-yield varieties of maize, sorghum, and millet were also developed. William Gaud, "The Green Revolution: Accomplishments and Apprehensions," AgBioWorld, March 8, 1968, www.agbioworld.org/biotech-info/topics/borlaug/borlaug-green.html.

[28] Deepali Singhal Kohli and Nirvikar Singh, "The Green Revolution in Punjab, India: The Economics of Technological Change" (paper presented at Agriculture of the Punjab conference at the Southern Asian Institute, Columbia University, April 1, 1995; revised in September 1997), 2, http://people.ucsc.edu/~boxjenk/greenrev.pdf.

[29]Daniel Zwerdling, "India's Farming 'Revolution' Heading for Collapse," *All Things Considered*, April 13, 2009, www.npr.org/templates/story/story.php?storyId=102893816.

[30]See Daniel Pepper, "The Toxic Consequences of the Green Revolution," *U.S. News and World Report*, July 7, 2008, www.usnews.com/news/world/articles/2008/07/07/the-toxic-consequences-of-the-green-revolution.

[31]Muir, "Consequences for Organic Matter in Soils."

[32]Columbia Water Center, "Punjab, India," Columbia University Earth Institute, http://water.columbia.edu/research-projects/india/punjab-india/.

[33]Cited from Zwerdling, "India's Farming 'Revolution' Heading for Collapse."

[34]For the latest UN data on the population of India see "India Population 2019, Most Populated States," www.indiapopulation2019.in. I am grateful to say that with the help of the Columbia Water Center of Columbia University a small and inexpensive soil-moisture measurement tool known as a "tensiometer" is helping small farmers reduce their water usage significantly (see Columbia Water Center, "Punjab, India," and Lakis Polycarpou, "'Small Is Also Beautiful'—Appropriate Technology Cuts Rice Farmers' Water Use by 30 Percent in Punjab India," *State of the Planet* (blog), Columbia University Earth Institute, November 17, 2010, https://blogs.ei.columbia.edu/2010/11/17/%e2%80%9csmall-is-also-beautiful%e2%80%9d-%e2%80%93-appropriate-technology-cuts-rice-famers%e2%80%99-water-use-by-30-percent-in-punjab-india/). We can also be encouraged that organic farming efforts in India are offering some hope to this region. See Yogita Limaye, "Will Organic Revolution Boost Farming in India?," *BBC News*, September 24, 2018, www.bbc.com/news/av/business-45605018/will-organic-revolution-boost-farming-in-india.

[35]Information cited from "Mississippi Agriculture Overview," Mississippi Department of Agriculture and Commerce, www.mdac.ms.gov/agency-info/mississippi-agriculture-snapshot/. See also the Farm Families of Mississippi website, which states that "agriculture is Mississippi's number one industry, employing roughly 260,000 people—17% of the state's workforce—either directly or indirectly. Agriculture contributes $7.4 billion in income—a full 22% of the state's total.... Plus, [there is] an estimated $2.7 billion annual economic impact from hunting, fishing, and other natural resource related enterprises" ("Mississippi's Heritage. Mississippi's Future," Farm Families of Mississippi, https://growingmississippi.org/agriculture-in-mississippi/).

[36]Michael Pollan, "An Open Letter to the Next Farmer in Chief," *New York Times Magazine*, October 12, 2008, www.nytimes.com/2008/10/12/magazine/12policy-t.html. According to Tom Starrs from the Center for Ecoliteracy, "Overall, about 15 percent of U.S. energy goes to supplying Americans with food, split roughly equally between crop and livestock production and food processing and packaging. David Pimentel, a professor of ecology and agricultural science at Cornell University, has estimated that if all humanity ate the way Americans eat, we would exhaust all known fossil fuel reserves in just seven years" (Tom Starrs, "Fossil Food: Consuming Our Future," Center for Ecoliteracy, June 29, 2009, www.ecoliteracy.org/article/fossil-food-consuming-our-future).

3 THE DOMESTIC CREATURES ENTRUSTED TO ʾĀDĀM

[1]See Sandra L. Richter, *The Epic of Eden: A Christian Entry into the Old Testament* (Downers Grove, IL: IVP Academic, 2008), 69-91, for a full treatment of the concept of covenant.

[2] Henri Blocher, *In the Beginning: The Opening Chapters of Genesis* (Downers Grove, IL: InterVarsity Press, 1984), 57.

[3] Abraham Heschel, *The Sabbath: Its Meaning for Modern Man* (New York: Farrar, Straus & Giroux, 1951), 6.

[4] "By the seventh millennium BC goats and sheep were domesticated in Mesopotamia, and 'pastoralism' has remained central to the economies in the Fertile Crescent up until the modern era. Ancient Mesopotamian states managed flocks numbering in the tens and even hundreds of thousands" (Timothy S. Laniak, *Shepherds After My Own Heart: Pastoral Traditions and Leadership in the Bible*, New Studies in Biblical Theology 20 [Downers Grove, IL: InterVarsity Press, 2006], 42-43).

[5] Oded Borowski, *Every Living Thing: The Daily Use of Animals in Ancient Israel* (Lanham, MD: AltaMira, 1999), 52-61.

[6] A thorough treatment of the textile industry in the ancient Near East can be found in Catherine Breniquet and Cécile Michel, eds., *Wool Economy in the Ancient Near East and the Aegean: From the Beginnings of Sheep Husbandry to Institutional Textile Industry*, Ancient Textile Series 17 (Oxford: Oxbow, 2014).

[7] See Timothy S. Laniak, *While Shepherds Watch Their Flocks: Forty Daily Reflections on Biblical Leadership* (n.p.: ShepherdLeader Publications, 2007), 109-10; Laniak, *Shepherds After My Own Heart*, 46-51. See as well Philip J. King and Lawrence E. Stager, *Life in Biblical Israel* (Louisville, KY: Westminster John Knox, 2001), 112-14; Borowski, *Every Living Thing*, 41-70. And see "Breeds of Livestock—Awassi Sheep," Department of Animal Science, Oklahoma State University, www.ansi.okstate.edu/breeds/sheep/awassi, and "Breeds of Livestock—Anatolian Black Goats," Department of Animal Science, Oklahoma State University, www.ansi.okstate.edu/breeds/goats/anatolianblack, for images and current animal husbandry practices involving these animals.

[8] Borowski, *Every Living Thing*, 61-65; cf. Virginia A. Finch et al., "Why Black Goats in Hot Deserts? Effects of Coat Color on Heat Exchanges of Wild and Domestic Goats," *Physiological Zoology* 53, no. 1 (1980): 19.

[9] Barley and wheat were well-suited to Levantine conditions, and their abundance in Canaan is frequently mentioned in Egyptian literature. In his sixteen campaigns into Canaan, Thutmose III records grain looted from Canaan several times, and in his fifth campaign he speaks of the grain on the threshing floors as "more plentiful than the sands of the shore" (James B. Pritchard, ed., *Ancient Near Eastern Texts Relating to the Old Testament*, 3rd ed. [Princeton, NJ: Princeton University Press, 1969], 239; cf. Oded Borowski, *Agriculture in Iron Age Israel* [Boston: American Schools of Oriental Research, 2002], 3-5). Baruch Rosen reports that storage silos and carbonized seeds have been found "in almost every Iron Age I site," and the tools and installations necessary to the sowing, reaping, threshing, winnowing, and storing of these grains have been the source of a bonanza of recovered material culture ("Subsistence Economy in Iron Age I," in *From Nomadism to Monarchy: Archaeological and Historical Aspects of Early Israel*, ed. Israel Finkelstein and Nadav Na'aman [Jerusalem: Israel Exploration Society, 1994], 343; cf. Jane Renfrew, "Vegetables in the Ancient Near Eastern Diet," in *Civilizations of the Ancient Near East*, ed. Jack M. Sasson [1995; repr. in 2 vols., Peabody, MA: Hendrickson, 2006], 1:195). Moreover, the barley and wheat harvests were central to the Israelite cultic calendar (Borowski, *Agriculture in Iron Age Israel*, 33, cf. 47-69).

[10] See Sandra L. Richter, "The Question of Provenance and the Economics of Deuteronomy," *Journal for the Study of the Old Testament* 42 (2017): 23-50.

[11] See Sandra L. Richter, "Environmental Law in Deuteronomy: One Lens on a Biblical Theology of Creation Care," *Bulletin for Biblical Research* 20 (2010): 371-72.

[12] Rosen, "Subsistence Economy in Iron Age I," 348-49; cf. Rosen's more detailed presentation of the same data in "Subsistence Economy of Stratum II," in *'Izbet Ṣarṭah: An Early Iron Age Site near Rosh Ha'ayin, Israel*, ed. Israel Finkelstein, BAR International Series 299 (Oxford: B.A.R., 1986), 156-85.

[13] This statistic emerges from a conservative estimate of the ancient working bovine at 600-700 pounds (275-320 kilograms). Such an animal should be able to comfortably consume as many as four kilos of wheat during a day of labor (personal communication, Ryan Strebeck of the Strebeck Family Ranch of the Curry and Roosevelt Counties in New Mexico and Elk City, Kansas, October 29, 2008). This intake is reasoned off the daily dry ration of a mature Angus steer in a feedlot—a ration based on a percentage of body weight, which in the arid conditions of Texas and New Mexico averages 800 pounds. In comparison, the weight of Boran cattle in East Africa under less than favorable conditions is 350-400 pounds (160-180 kilograms), while its well-fed counterpart would average 990 pounds (450 kilograms; Gudrun Dahl and Anders Hjort, *Having Herds: Pastoral Herd Growth and Household Economy*, Stockholm Studies in Social Anthropology 2 [Stockholm: Dept. of Social Anthropology, University of Stockholm, 1976], 163-67). Nimrod Marom of the Laboratory of Archaeozoology at the University of Haifa states that ancient bovine weight is difficult to delineate but estimates 400 kilograms +/- 50 kilograms (personal communication, Nimrod Marom, October 28, 2008). Marom's estimate would increase caloric intake and therefore increase the sacrifice of the Israelite farmer when choosing not to muzzle his ox (Richter, "Environmental Law in Deuteronomy," 371-72; cf. Rosen, who estimates a higher intake of 5 kilograms [11 pounds] per day, "Subsistence Economy of Stratum II," 156-85).

[14] "Wean to finish" buildings are floored with slatted plastic, concrete, or metal that allows waste to be collected below the confinement pens. This waste is then flushed into open-air pits known as "waste lagoons." Waste-lagoon management has received a lot of attention in the press over the last decade. For a summary see Matthew Scully, *Dominion: The Power of Man, the Suffering of Animals, and the Call to Mercy* (New York: St. Martin's Griffin, 2002), 249, and Yanxia Li, Zhonghong Wu, Glen Broderick, and Brian Holmes, "Rapid Assessment of Feed and Manure Nutrient Management on Confinement Dairy Farms," *Nutrient Cycling in Agroecosystems* 82, no. 2 (2008): 107.

[15] Most conventional hogs in this country are Duroc breed, going to slaughter at 220-240 pounds (personal communication, Ann Bell Stone, October 28, 2008; cf. Scully, *Dominion*, 252).

[16] According to the USDA as of December 1, 2018, there were 74.6 million hogs and pigs on US farms. "United States Hog Inventory Up 2 Percent," www.nass.usda.gov/Newsroom/2018/12-20-2018.php, accessed June 16, 2019.

[17] According to the National Hog Farmer website, many producers have adopted W-F because it simplifies pig flow and recordkeeping, it saves as much as $1 per pig, and it retrieves the lost day of growth produced by moving the pigs from the "nursery" pen

to the "finish" pen (Mike Brumm, "Wean-to-Finish Systems: An Overview," National Hog Farmer, October 1, 1999, www.nationalhogfarmer.com/mag/farming_weanto finish_systems_overview).

[18] Scully, *Dominion*, 248-61. See Terry Feldmann, "Equipment, Facility Designs," National Hog Farmer, October 1, 1999, www.nationalhogfarmer.com/mag/farming _equipment_facility_designs.

[19] John Webster, *Animal Welfare: Limping Towards Eden; A Practical Approach to Redressing the Problem of Our Dominion over the Animals* (Oxford: Blackwell, 2005), 110-19; cf. J. P. Tillon and F. Madec, "Diseases Affecting Confined Sows: Data from Epidemiological Observations," *Annales de Recherches Vétérinaires, INRA Editions* 15, no. 2 (1984): 195-99; Nicoline M. Soede and Bas Kemp, "Housing Systems in Pig Husbandry Aimed at Welfare; Consequences for Fertility," an abbreviated and updated version of "Reproductive Issues in Welfare Friendly Housing Systems in Pig Husbandry: A Review," *Reproduction in Domestic Animals* 47 (Supplement 5, 2012): 51-57; D. M. Broom, M. T. Mendl, and A. J. Zanella, "A Comparison of the Welfare of Sows in Different Housing Conditions," *Animal Science* 61 (1995): 369-85.

[20] "An HSUS Report: Welfare Issues with Gestation Crates for Pregnant Sows," The Humane Society of the United States, February 2013, www.humanesociety.org/sites/default/files /docs/hsus-report-gestation-crates-for-pregnant-sows.pdf; *Quarterly Hogs and Pigs*, National Agricultural Statistics Service, July 27, 2019, www.nass.usda.gov/Publications /Todays_Reports/reports/hgpg0619.pdf; Roberta Lee, "Summer Fun, but Not for Pigs: The Horror of Gestation Crates and Life in a Factory Farm," *Huffington Post*, July 16, 2015, www.huffpost.com/entry/summer-fun-but-not-for-pi_b_7759466. Scully, *Dominion*, 247-86. For more information and images of gestation crates in use, see www.farmsanctuary .org/learn/factory-farming/pigs-used-for-pork. Reading of the standard treatment that these animals endure, one cannot help but think of Ezekiel's outcry against the shepherds of Israel in Ezek 34:3-4: "You eat the fat and clothe yourselves with the wool, you slaughter the fat *sheep* without feeding the flock. Those who are sickly you have not strengthened, the diseased you have not healed, the broken [i.e., those with broken bones] you have not bound up, the scattered you have not brought back, nor have you sought for the lost; but with force and with severity you have dominated them" (NASB).

[21] See Scully, *Dominion*, 261-66; cf. Danny Na and Tom Polansek, "U.S. Hogs Fed Pig Remains, Manure to Fend Off Deadly Virus Return," *Scientific American*, www .scientificamerican.com/article/u-s-hogs-fed-pig-remains-manure-to-fend-off-deadly -virus-return/; and Eliza Barclay's NPR report, "'Piglet Smoothie' Fed to Sows to Prevent Disease; Activists Outraged," *The Salt*, February 20, 2014, www.npr.org/sections /thesalt/2014/02/20/280183550/piglet-smoothie-fed-to-sows-to-prevent-disease -activists-outraged.

[22] The good news is that gestation crates were fully outlawed in the European Union in 2013. California and Massachusetts have outlawed them. But other states trail behind. Why? One reason is that pork production is big business. In North Carolina this industry has a total economic impact estimated at $9 billion, directly contributing over $2.5 billion to the North Carolina state economy. The tax revenue toward local schools and infrastructure and business is significant enough that most are willing to look the other way. See "The Facts," North Carolina Farm Families, https://ncfarmfamilies.com/thefacts/.

[23] "Overview of U.S. Livestock, Poultry, and Aquaculture Production in 2010 and Statistics on Major Commodities," Animal and Plant Health Inspection Service of the USDA, www.aphis.usda.gov/animal_health/nahms/downloads/Demographics2010_rev.pdf, 12-13.

[24] "Cage-Free vs. Battery-Cage Eggs," The Humane Society of the United States, www.humanesociety.org/resources/cage-free-vs-battery-cage-eggs.

[25] "Cage-Free vs. Battery-Cage Eggs."

[26] "Animals on Factory Farms," American Society for the Prevention of Cruelty to Animals (ASPCA), www.aspca.org/animal-cruelty/farm-animal-welfare/animals-factory-farms. See Webster, *Animal Welfare*, 120-25.

[27] See the ASPCA report, "A Growing Problem: Selective Breeding in the Chicken Industry; The Case for Slower Growth," ASPCA, 3, www.aspca.org/sites/default/files/chix_white_paper_nov2015_lores.pdf.

[28] "Growing Problem," 3.

[29] "Growing Problem," 3.

[30] In June of 2017 Pilgrim's Pride was the subject of a major undercover investigation regarding animal abuse. The video exposé is available on the Humane Society's website. It is very difficult to watch. Charges for cruelty to animals were filed with the Madison County Sheriff's Department in Danielsville, Georgia, and the Titus County Sheriff's Office in Mt. Pleasant, Texas. See "Shocking Animal Abuse Uncovered at Country's Second Largest Chicken Producer," press release, Humane Society of the United States, June 27, 2017, www.humanesociety.org/news/shocking-animal-abuse-uncovered-countrys-second-largest-chicken-producer.

[31] "Growing Problem," 4. See Tyson Foods, "Contract Poultry Farming," www.tysonfoods.com/who-we-are/our-partners/farmers/contract-poultry-farming.

[32] "Growing Problem," 2.

[33] "Growing Problem," 5.

[34] The enormous energy consumption required for mass-confinement animal husbandry is an interesting part of this equation. In his recent "An Open Letter to the Next Farmer in Chief," *New York Times Magazine*, October 12, 2008, www.nytimes.com/2008/10/12/magazine/12policy-t.html, Michael Pollan states that the industry currently consumes 19 percent of the annual national consumption of fuel. In 1940 each calorie of fossil fuel produced 2.3 calories of food, whereas the current ratio is 10 calories of fossil fuel to every 1 calorie of food.

[35] Eric Schlosser, "Cheap Food Nation," *Sierra*, November/December 2006, 36-39, https://vault.sierraclub.org/sierra/200611/cheapfood.asp.

[36] See Jacob Milgrom, *Leviticus: A Book of Ritual and Ethics*, Continental Commentaries (Minneapolis: Fortress, 2004), 184-92. Leviticus 17:4 states that "bloodguiltiness" (i.e., murder) will be on the person who slaughters without taking the animal before the priest.

[37] Cf. Miller, *Religion of Ancient Israel*, 126.

[38] Milgrom, *Leviticus*, 106.

[39] Milgrom, *Leviticus*, 106.

[40] The live grinding of unwanted male chicks is one of the more gruesome practices of the poultry agribusiness. Gruesome video of this practice is available on the PETA website, www.peta.org/students/missions/male-chicks-ground-up-alive/. There is some hope,

as reported by Maryn McKenna, "By 2020, Male Chicks May Avoid Death by Grinder," *National Geographic*, June 13, 2016, www.nationalgeographic.com/people-and-culture/food/the-plate/2016/06/by-2020--male-chicks-could-avoid-death-by-grinder/.

[41] Interview with reporter Joby Warrick in 2001, found in Scully, *Dominion*, 284.

[42] Scully, *Dominion*, 284.

[43] In a PBS interview with *Frontline* titled "Inside the Slaughterhouse," Eric Schlosser (author of *Fast Food Nation*) commented on the overwhelmingly high turnover rates in American slaughterhouses (75-100 percent per year). He stated, "Work in a slaughterhouse has changed enormously in the last 25 years. It's always been a difficult job. It's always been a dangerous job. But up until recently, this was a job that had good pay, had good benefits, and you had a very stable work force. In the early 1970s, meatpacking had one of the lowest turnover rates of any industrial job in America. It was like being an autoworker. Then they cut wages, they cut benefits, broke unions. And now it has one of the highest turnover rates of any industrial job." *Frontline*, "Inside the Slaughterhouse," www.pbs.org/wgbh/pages/frontline/shows/meat/slaughter/slaughterhouse.html.

[44] See "Factory Farms," ASPCA, www.aspca.org/animal-cruelty/farm-animal-welfare.

[45] See "What Is Ag-Gag Legislation?," ASPCA, www.aspca.org/animal-protection/public-policy/what-ag-gag-legislation.

[46] See Luke Runyon, "Judge Strikes Down Idaho 'Ag-Gag' Law, Raising Questions for Other States," NPR, August 4, 2015, www.npr.org/sections/thesalt/2015/08/04/429345939/idaho-strikes-down-ag-gag-law-raising-questions-for-other-states.

[47] Leighton Akio Woodhouse, "Charged with the Crime of Filming a Slaughterhouse," *The Nation*, July 31, 2013, www.thenation.com/article/charged-crime-filming-slaughterhouse/. The video footage of the confrontation between Amy Meyer and the local police can be viewed at this website.

[48] Woodhouse, "Filming a Slaughterhouse."

[49] Woodhouse, "Filming a Slaughterhouse."

[50] Woodhouse, "Filming a Slaughterhouse."

[51] Susan Allen, "4 Modern Milking Parlor Designs," Dairy Discovery Zone, December 13, 2017, www.dairydiscoveryzone.com/blog/4-modern-milking-parlor-designs.

[52] Jim Goodman, "Dairy Farming Is Dying. After 40 Years, I'm Done," *Washington Post*, December 21, 2018, www.washingtonpost.com/outlook/dairy-farming-is-dying-after-40-years-im-out/2018/12/21/79cd63e4-0314-11e9-b6a9-0aa5c2fcc9e4_story.html?utm_term=.d4192c32708c.

[53] Goodman, "Dairy Farming Is Dying."

[54] Goodman, "Dairy Farming Is Dying."

[55] Goodman, "Dairy Farming Is Dying."

[56] Myrto Theocharous, "Becoming a Refuge: Sex Trafficking and the People of God," *Journal of the Evangelical Theological Society* 59 (2016): 318.

[57] Theocharous, "Becoming a Refuge," 318.

4 THE WILD CREATURES ENTRUSTED TO ʾĀDĀM

A particularly interesting expression of Jewish environmentalism is the current call to an "eco-kosher" lifestyle. Whereas "kosher" is defined as "fit, proper or in accordance with the religious law" and has historically been applied only to dietary laws, the current call is to a lifestyle that is "proper" because it protects the earth (see Arthur Waskow, "What Is Eco-Kosher?," in Gottlieb, *This Sacred Earth*, 297-300; Rabbi Hayim Halevy Donin, *To Be a Jew: A Guide to Jewish Observance in Contemporary Life* [New York: Basic Books, 1972], 97).

[1] The American Farmland Trust estimates that, between 1992 and 2012, we lost nearly 31 million acres of farmland: 175 acres an hour, 3 acres every minute ("Farms Under Threat," American Farmland Trust, https://www.farmland.org/initiatives/farms-under-threat).

The National Resource Defense Council published a helpful treatment of the topic; see Deron Lovaas, "Measuring Suburban Sprawl," National Resource Defense Council, April 2, 2014, https://www.nrdc.org/experts/deron-lovaas/measuring-suburban-sprawl.

There has also been a dramatic shift in recent years regarding federal protection of wildlife and their habitat. According to one report, "Lands proposed as critical habitat by biologists were reduced by one-third; 69 percent of those reductions were based on economic factors, up from fewer than 1 percent in 2001" (Felicity Barringer, "Endangered Species Act Faces Broad New Challenges," *New York Times*, June 26, 2005, www.nytimes.com/2005/06/26/politics/endangered-species-act-faces-broad-new-challenges.html). Daniel Glick reports that President Bush's Executive Order 13212, "Actions to Expedite Energy-Related Projects," on May 18, 2001, may be characterized as "the most far-ranging and destructive swipes of a pen that a president has inflicted on federally administered public lands" ("Putting the 'Public' Back in Public Lands: An Open Letter to the Next President," *National Wildlife*, October/November 2008, 26). As a result, the number of annual approvals for drilling permits in the Rocky Mountain region increased 125 percent between 2001 and 2007, jumping from 1,500 in 2000 to nearly 6,500 in 2007 (Glick, "Putting the 'Public' Back," 27).

[2] Whereas the historically expected rate of species extinction should be about one species in a million annually, "studies of various organisms (birds, mussels, fish, and plants) show that these groups are now disappearing more than 100 times faster, and in some cases up to 1,000 times faster, than the background rate. Even worse, the number of species currently threatened with extinction far exceeds those recently lost, bringing future extinction estimates to potentially 10,000 times the 'normal' rate" (Kyle S. Van Houtan, "Extinction and Its Causes," *Creation Care*, Fall 2008, 15).

[3] Daniel Swartz, "Jews, Jewish Texts, and Nature: A Brief History," in *This Sacred Earth: Religion, Nature, Environment*, ed. Roger S. Gottlieb (New York: Routledge, 1996), 92-109 (originally published in *To Till and to Tend: A Guide to Jewish Environmental Study and Action* [New York: The Coalition on the Environment and Jewish Life, 1995]).

[4] The Israeli Nature and Parks Authority is working hard to reintroduce a number of these species, although habitat destruction continues to be a major roadblock (see Noam Kirshenbaum, *Mammals of Israel: A Pocket Guide to Mammals and Their Tracks*, Nature in Israel series [The National Parks Authority, 2005]).

[5] Nathan MacDonald, *What Did the Ancient Israelites Eat? Diet in Biblical Times* (Grand Rapids: Eerdmans, 2008), 34.

[6] US Armed Forces Medical Intelligence Center, United States Defense Intelligence Agency Deputy Directorate for Scientific and Technical Intelligence, *Venomous Snakes of the Middle East Identification Guide* (Fort Detrick, Frederick, MD: Armed Forces Medical Intelligence Center, 1991), Department of Defense Intelligence Document DST-1810S-469-91.

[7] Oded Borowski, *Every Living Thing: The Daily Use of Animals in Ancient Israel* (Lanham, MD: AltaMira, 1999), 196-205.

[8] See Carol Meyers, "The Family in Early Israel," in *Families in Ancient Israel*, ed. Leo G. Perdue et al., The Family, Religion, and Culture (Louisville, KY: Westminster John Knox, 1997), 1-47, and Avraham Faust, *The Archaeology of Israelite Society in Iron Age II* (Winona Lake, IN: Eisenbrauns, 2012), 7-27, 255.

[9] See Sandra L. Richter, "The Question of Provenance and the Economics of Deuteronomy," *Journal for the Study of the Old Testament* 42 (2017): 23-50; cf. Deut 7:13; 11:14; 12:17; 14:23; 16:9; 18:4; 23:25; 28:51; 33:28.

[10] See David C. Hopkins, "The Dynamics of Agriculture in Monarchical Israel," in *Society of Biblical Literature 1983 Seminar Papers*, ed. Kent Harold Richards (Chico, CA: Scholars Press, 1983), 177-93; Oded Borowski, *Daily Life in Biblical Times*, Archaeology and Biblical Studies 5 (Atlanta: Society of Biblical Literature, 2003), 13-42; and Avraham Faust, "Cities, Villages, and Farmsteads: The Landscape of Leviticus 25.29-31," in *Exploring the Longue Durée: Essays in Honor of Lawrence E. Stager*, ed. J. David Schloen (Winona Lake, IN: Eisenbrauns, 2009), 106.

[11] See Richter, "Question of Provenance," 26-29.

[12] Jacob Wright, "Warfare and Wanton Destruction," *Journal of Biblical Literature* 127, no. 3 (2008): 453; cf. Richard Nelson, *Deuteronomy*, Old Testament Library (Louisville, KY: Westminster John Knox, 2002), 268.

[13] Nelson, *Deuteronomy*, 337.

[14] Duane L. Christensen, *Deuteronomy 21:10–34:12*, Word Biblical Commentary (Nashville: Thomas Nelson, 2002), 500; Jeffrey Tigay, *Deuteronomy*, JPS Torah Commentary (Philadelphia: The Jewish Publication Society, 1996), 201. Sasson's interpretation of the Deuteronomic law regarding boiling a kid in its mother's milk echoes the same concern (Jack Sasson, "Should Cheeseburgers Be Kosher?" *Biblical Research* 19 [2008]: 40-43, 50-51).

[15] Sandra L. Richter, "Eighth-Century Issues: The World of Jeroboam II, the Fall of Samaria, and the Reign of Hezekiah," in *Ancient Israel's History: An Introduction to Issues and Sources*, ed. Bill T. Arnold and Richard S. Hess (Grand Rapids: Baker Academic, 2014), 337-40.

[16] Richter, "Question of Provenance," 35-38.

[17] See Benedikt Otzen, "Israel Under the Assyrians," in *Power and Propaganda, Mesopotamia 7*, ed. M. T. Larsen (Copenhagen: Akademisk Forlag, 1979), 251-56.

[18] Richter, "Eighth-Century Issues," 337.

[19] In these reliefs from the northern palace at Kuyûnjik servants of the king are pictured carrying dead lions, a hare, a bird, and birds' nests back from the royal hunt. This panel is part of the larger Lion Hunt Relief exhibit at the British Museum, in which dozens of hunt scenes are depicted. The graphic celebration of the slaughter of wild creatures by the king is understood to be a public promulgation of the legitimacy of his rule (C. J. Gadd, *The Assyrian Sculptures*, The British Museum [London: Harrison & Sons, 1934], 72-73; cf. Wright, "Warfare and Wanton Destruction," 454, fig. 4).

[20] "Bottomland forests represent a transition between drier upland hardwood forest and very wet river floodplain and wetland forests. While trees and plants in the bottomland hardwood forest cannot tolerate long periods of flooding (as in a swamp), they are

flooded periodically when water levels rise" ("Bottomland Hardwoods," University of Florida Forest Resources and Conservation paper, 2, www.sfrc.ufl.edu/extension/4h/ecosystems/bottomland_hardwoods/bottomland_hardwoods_description.pdf). These highly diverse ecosystems can support two to five times the number of species as a pine or upland hardwood forest, they are critical to breeding fish and birds, and they serve an important function in cleansing waterways ("Bottomland Hardwoods," 2-7).

[21]James A. Allen, "Reforestation of Bottomland Hardwoods and the Issue of Woody Species Diversity," *Restoration Ecology* 5, no. 2 (June 1997): 125-34; cf. Brad Young, "Black Bears in Mississippi Past and Present," *Wildlife Issues*, Fall/Winter 2004, www.mdwfp.com/media/3306/ms_black_bear_wildlife_issues_2004.pdf.

[22]One of the most devastating results of urban sprawl in the United States has been the destruction of wetlands. Older statistics from the US Fish and Wildlife Service report that roughly 58,500 acres of wetlands are being destroyed annually, according to the most recent US Geological Survey study. The great state of Louisiana has lost 2,000 square miles of wetlands in the last eighty years; see Jonathan Hahn, "The Tragedy and Wonder of Louisiana's Wetlands: Photographer Ben Depp Uses a Paraglider to Document a Fading Ecosystem," Sierra Club, www.sierraclub.org/sierra/slideshow/tragedy-and-wonder-louisiana-s-wetlands. Yet wetlands serve an array of critical roles in the survival of every species on this planet. For a focused introduction to this far-ranging problem see *Audubon Magazine*'s special May 2006 issue titled *America's River*, which provides an exposé of the broad impact of the long-term abuse of the mighty Mississippi.

[23]Teddy Roosevelt was known as a "big game hunter," but even he would not take out a beaten and tied black bear secured for him by an aid near Onward, Mississippi, on November 14, 1902. When Clifford Berryman, a political cartoonist, caricatured the event in the *Washington Post* on November 16 of the same year, Morris Michtom of Brooklyn decided to create and dedicate a stuffed toy bear to the president who would not participate in canned hunting. Thus the "Teddy Bear," and eventually the Ideal Toy Company, was born (see among other sources for this legendary tale, "The Story of the Teddy Bear," National Park Service, last updated February 16, 2019, www.nps.gov/thrb/learn/historyculture/storyofteddybear.htm).

[24]Stephanie L. Simek et al., "History and Status of the American Black Bear in Mississippi," *Ursus* 23, no. 2 (2012): 159-67.

[25]"Range, Movements and Sightings," Mississippi Wildlife, Fisheries, and Parks, www.mdwfp.com/wildlife-hunting/black-bear-program/mississippi-black-bear-ecology/range-movements-and-sightings/. See the Wetland Reserve Program (WRP) and the Conservation Reserve Program (CRP) in Mississippi for details. These groups are attempting to replant thousands of acres that were stripped for pine production in order to restore bottomland hardwood species and thereby increase both habitat and corridors linking isolated patches of habitat throughout the Delta. Bonnie A. Coblentz, "Black Bear Numbers Rising in Mississippi," Mississippi State University Extension, July 28, 2005, http://extension.msstate.edu/news/feature-story/2005/black-bear-numbers-rising-mississippi. This report doubled the 2005 cooperative "den check" report (by Entergy, the Mississippi and Louisiana Departments of Wildlife Protection, Louisiana State University, and US Fish and Wildlife Service biologists) that documented the first birth of bear cubs in Mississippi in over forty years. (One breeding female located

in Wilkinson County with five healthy cubs [Simek, "History and Status," 159-67].)
[26] Allen, "Reforestation," 130.

5 ENVIRONMENTAL TERRORISM

[1] Jacob Wright, "Warfare and Wanton Destruction," *Journal of Biblical Literature* 127, no. 3 (2008): 453; cf. Richard Nelson, *Deuteronomy*, Old Testament Library (Louisville, KY: Westminster John Knox, 2002), 268.

[2] Nelson, *Deuteronomy*, 337.

[3] Oded Borowski, *Agriculture in Iron Age Israel* (Boston: American Schools of Oriental Research, 2002), 100-133.

[4] Philip J. King and Lawrence E. Stager, *Life in Biblical Israel* (Louisville, KY: Westminster John Knox, 2001), 96.

[5] Sandra L. Richter, "Environmental Law in Deuteronomy: One Lens on a Biblical Theology of Creation Care," *Bulletin for Biblical Research* 20 (2010): 342-44; cf. Steven W. Cole, "The Destruction of Orchards in Assyrian Warfare," in *Assyria 1995: Proceedings of the 10th Anniversary Symposium of the Neo-Assyrian Text Corpus Project Helsinki, September 7-11, 1995*, ed. S. Parpola and R. M. Whiting (Helsinki: The Neo-Assyrian Text Corpus Project, 1997), 30. Jacob Wright identifies the production lifespan of the olive tree in centuries, and that of the date palm at over one hundred years ("Warfare and Wanton Destruction," 434).

[6] Cf. Jeremy Smoak, "Building Houses and Planting Vineyards: The Early Inner-Biblical Discourse on an Ancient Israelite Wartime Curse," *Journal of Biblical Literature* 127, no. 1 (2008): 19-35; Wright, "Warfare and Wanton Destruction"; Cole, "Destruction of Orchards," 29-40.

[7] Smoak, "Building Houses," 22; cf. Daniel David Luckenbill, ed., *Ancient Records of Assyria and Babylonia* (Chicago: University of Chicago Press, 1926–1927; repr., New York: Greenwood, 1968), 2:87, text 161.

[8] Smoak, "Building Houses," 21; cf. Grant Frame, ed., *Rulers of Babylonia from the Second Dynasty of Isin to the End of Assyrian Domination (1157–612 BC)*, The Royal Inscriptions of Mesopotamia, Babylonian Periods 2 (Toronto: University of Toronto Press, 1995), 295, "Assyrian troops cutting down date palms at the siege of Dilbat."

[9] Cole, "Destruction of Orchards," 29-40.

[10] Michael Hasel, *Military Practice and Polemic: Israel's Laws of Warfare in Near Eastern Perspective* (Berrien Springs, MI: Andrews University Press, 2005), 102-13. See Jacob Wright's review in *Journal of Biblical Literature* 125, no. 3 (2006): 577.

[11] Aren M. Maeir, Oren Ackermann, and Hendrik J. Bruins, "The Ecological Consequences of a Siege: A Marginal Note on Deuteronomy 20:19-20," in *Confronting the Past: Archaeological and Historical Essays on Ancient Israel in Honor of William G. Dever*, ed. Seymour Gitin, J. Edward Wright, and J. P Dessel (Winona Lake, IN: Eisenbrauns, 2006), 239-42. Here an enormous siege trench and berm have been excavated. These have been dated by both material culture and radiometric data to the Iron IIA period. Evidence indicates that the ecosystem surrounding Tel es-Sâfi/Gath "was used far beyond its maximum capacity" during this encounter, that the surrounding area was "probably denuded" of all vegetation, and that the topography of the site was permanently altered due to the siege works system and severe erosion (Maeir, Ackermann, and Bruins,

"Ecological Consequences," 242). If indeed these findings prove to be a siege trench and berm (see David Ussishkin's challenge in "On the So-Called Aramaean 'Siege Trench' in Tell eṣ-Ṣafi, Ancient Gath," *Israel Exploration Journal* 59, no. 2 [2009]: 137-57), this site provides the environmental impact of such methods in living color.

[12] Hasel, *Military Practice and Polemic*, 35.

[13] Geoffrey C. Ward and Ken Burns, *The Vietnam War: An Intimate History* (New York: Knopf, 2017), 158, 406. See as well Bao Ninh, a member of the Vietnam People's Army, speaking of his experience with "the dreadful nightmare of dioxin," which among all the horrors of war "is what comes back to me most often and disturbs my sleep" (Ward and Burns, *Vietnam War*, 461-63). See as well "Agent Orange," History, August 2, 2011, www.history.com/topics/vietnam-war/agent-orange-1.

[14] "Agent Orange," Military Wikia, https://military.wikia.org/wiki/Agent_Orange.

[15] "Agent Orange."

[16] Ben Stocking, "Agent Orange Still Haunts Vietnam, US," *Washington Post*, June 14, 2007, www.washingtonpost.com/wp-dyn/content/article/2007/06/14/AR2007061401077.html?noredirect=on.

[17] See the excellently cited synopsis, "Agent Orange," Wikipedia, http://en.wikipedia.org/wiki/Agent_Orange.

[18] Ash Anand, "Vietnam's Horrific Legacy: The Children of Agent Orange," *News Corp Australia*, May 25, 2015, www.news.com.au/world/asia/vietnams-horrific-legacy-the-children-of-agent-orange/news-story/c008ff36ee3e840b005405a55e21a3e1.

[19] Anand, "Vietnam's Horrific Legacy."

[20] Anand, "Vietnam's Horrific Legacy"; cf. Stocking, "Agent Orange Still Haunts Vietnam, US."

[21] The US Department of Veteran's Affairs lists Parkinson's; heart disease; and cancers of the lung, larynx, trachea, and prostate as "presumptive" diseases associated with exposure to Agent Orange ("Veterans' Diseases Associated with Agent Orange," US Department of Veterans Affairs, www.publichealth.va.gov/exposures/agentorange/diseases.asp). To read one of the most recent treatments of this issue, see Edwin Martini, *Agent Orange: History, Science, and the Politics of Uncertainty* (Boston: University of Massachusetts Press, 2012).

[22] See "Over $2.2 Billion in Retroactive Agent Orange Benefits Paid to 89,000 Vietnam Veterans and Survivors for Presumptive Conditions," US Department of Veteran's Affairs Office of Public and Intergovernmental Affairs, August 31, 2011, www.va.gov/opa/pressrel/pressrelease.cfm?id=2154; as well Kimberly Nicoletti, "The Aftermath of Agent Orange: Local Woman Forms Nonprofit to Aid Affected Children," Vietnam Agent Orange Relief and Responsibility Campaign, January 14, 2006, www.vn-agentorange.org/aspen_20060114.html. For a deeply painful photographic journal of the victims of the war, see Philip Jones Griffiths, *Agent Orange: Collateral Damage in Vietnam* (London: Trolley, 2004).

[23] Ralph Blumenthalmay, "Veterans Accept $180 Million Pact on Agent Orange," *New York Times*, June 16, 2019, www.nytimes.com/1984/05/08/nyregion/veterans-accept-180-million-pact-on-agent-orange.html.

[24] Fred A. Wilcox, "Toxic Agents: Agent Orange Exposure," in *The Oxford Companion to American Military History*, ed. John Whiteclay Chambers (Oxford: Oxford University Press, 1999), 725; Blumenthalmay, "Veterans Accept $180 Million Pact on Agent Orange."

[25]Wayne Dwernychuk, "Agent Orange and Dioxin Hot Spots in Vietnam," Persistent Organic Pollutants Toolkit, www.popstoolkit.com/about/articles/aodioxinhotspots vietnam.aspx.

6 THE WIDOW AND THE ORPHAN

[1]This last category is the Hebrew *gēr*, probably best understood in our world as an "immigrant" or a "refugee."

[2]For a full discussion of Israel's tribal culture and biblical examples of the same, see Sandra L. Richter, *The Epic of Eden: A Christian Entry into the Old Testament* (Downers Grove, IL: IVP Academic, 2008), 21-46. Cf. Marshall D. Sahlins's classic piece *Tribesmen* (Englewood Cliffs, NJ: Prentice Hall, 1968); Max Weber, "Bureaucracy" and "Patriarchalism and Patrimonialism," in *Economy and Society: An Outline of Interpretive Sociology*, ed. Guenther Roth and Claus Wittich (Berkeley: University of California Press, 1978), 2:956-1069; and Robert D. Miller's recent discussion of Israel's societal structure, "Complex Chiefdom Model," in *Chieftains of the Highland Clans: A History of Israel in the 12th and 11th Centuries B.C.* (Grand Rapids: Eerdmans, 2005), 6-28.

[3]See Richter, *Epic of Eden*, 25-38.

[4]Sahlins, *Tribesmen*, 15; cf. Miller, *Chieftains of the Highland Clans*, 6-28.

[5]Although the term *clan* is more familiar, Sahlins (*Tribesmen*, 15) clarifies that this layer is better described as "local lineages" that are gathered into village communities. Carol Meyers ("The Family in Early Israel," in *Families in Ancient Israel*, ed. Leo G. Perdue et al., The Family, Religion, and Culture [Louisville, KY: Westminster John Knox, 1997], 13) opts for the term "residential kinship group."

[6]Lawrence E. Stager, "Archaeology of the Family in Ancient Israel," *Bulletin of the American Schools of Oriental Research* 260 (1985): 20-22. This societal structure is reflected throughout the Old Testament, but it is particularly visible in the account of Josh 7:14-15, where Joshua identifies the individual who has violated the ban in the battle of Ai by using lots to first identify the guilty tribe, then the clan, then the household, and then the individual (Stager, "Archaeology of the Family in Ancient Israel," 22; cf. Philip J. King and Lawrence E. Stager, *Life in Biblical Israel* [Louisville, KY: Westminster John Knox, 2001], 36-38).

[7]King and Stager, *Life in Biblical Israel*, 39-40.

[8]Meyers comments that these family configurations tend to form "in situations in which labor requirements are so demanding that a residential group cannot survive at subsistence level without the productive labor of more than a conjugal pair and their children" ("Family in Early Israel," 18).

[9]For a full discussion of Israel's tribal culture and biblical examples of the same, see Richter, *Epic of Eden*, 21-46.

[10]For further reading see King and Stager, *Life in Biblical Israel*, 53-57.

[11]Meyers, "The Family in Early Israel," 3; cf. Oded Borowski, *Daily Life in Biblical Times*, Archaeology and Biblical Studies 5 (Atlanta: Society of Biblical Literature, 2003), 13-42; and David C. Hopkins, "Life on the Land: The Subsistence Struggles of Early Israel," *Biblical Archaeologist* 50 (1987): 178-91.

[12]Avraham Faust, "Cities, Villages, and Farmsteads: The Landscape of Leviticus 25.29-31," in *Exploring the Longue Durée: Essays in Honor of Lawrence E. Stager*, ed. J. David Schloen (Winona Lake, IN: Eisenbrauns, 2009), 106; cf. "Bethlehem," *NEAEHL* 1:203-8.

[13] If you are looking for a study on this classic tale, visit seedbed.com for my *Epic of Eden: Ruth* DVD-based curriculum.

[14] For further reading, see Christopher J. H. Wright, *God's People in God's Land: Family, Land, and Property in the Old Testament* (Grand Rapids: Eerdmans, 1990); Meyers, "Family in Early Israel," 19-21; Joseph Blenkinsopp, "The Family in First Temple Israel," in Perdue et al., *Families in Ancient Israel*, 54-56; and Borowski, *Daily Life*, 26-27.

[15] Meyers, "Family in Early Israel"; David C. Hopkins, *The Highlands of Canaan: Agricultural Life in the Early Iron Age*, Social World of Biblical Antiquity 3 (Decatur, GA: Almond, 1985); Hopkins, "Life on the Land," 178-91; and Aharon Sasson, *Animal Husbandry in Ancient Israel: A Zooarchaeological Perspective on Livestock Exploitation, Herd Management and Economic Strategies* (London: Equinox, 2010), 60-61 and 119-22.

[16] Oded Borowski (*Agriculture in Iron Age Israel* [Boston: American Schools of Oriental Research, 2002], 60) interprets this term ʿōmer as a stalk or bunch of wheat, or barley stalks not yet bound into sheaves.

[17] See Richter, "Question of Provenance," for a full discussion of Israel's evolving economy in the Iron Age.

[18] Excavations at Philistine Ekron have produced more than one hundred olive presses, and "archaeologists speculate that Ekron produced a thousand tons of oil annually, mostly for export" (King and Stager, *Life in Biblical Israel*, 96). Under Neo-Assyrian hegemony Ekron grew into the second-largest city in Judah, with the city becoming "the single largest production center in all of the premodern Middle East, reaching oil capacities many times beyond the local consumption" (Roger S. Nam, *Portrayals of Economic Exchange in the Book of Kings*, Biblical Interpretation Series 112 [Leiden: Brill, 2012], 129).

[19] See King and Stager, *Life in Biblical Israel*, 98.

[20] Borowski, *Agriculture in Iron Age Israel*, 102-14; cf. Shalom M. Paul and William G. Dever, *Biblical Archaeology* (New York: Quadrangle & New York Times, 1974), fig. 77.

[21] See Borowski, *Agriculture in Iron Age Israel*, 110.

[22] Meyers, "Family in Early Israel," 3; cf. Borowski, *Daily Life*, 13-42; and Hopkins, "Life on the Land."

[23] Norman Wirzba, "The Grace of Good Food and the Call to Good Farming," *Review and Expositor* 108 (Winter 2011): 61-71; cf. Wirzba, *The Paradise of God: Renewing Religion in an Ecological Age* (New York: Oxford University Press, 2003).

[24] Thorkild Jacobsen and Robert M. Adams, "Salt and Silt in Ancient Mesopotamian Agriculture," *Science* 128 (1958): 1251-58. Cf. Marvin A. Powell, "Salt, Seed, and Yields in Sumerian Agriculture: A Critique of the Theory of Progressive Salinization," *Zeitschrift für die Assyriologie* 75 (1985): 7-38.

[25] Wendell Berry, *The Unsettling of America: Culture and Agriculture*, 3rd ed. (San Francisco: Sierra Club Books, 1996); David R. Montgomery, *Dirt: The Erosion of Civilizations* (Berkeley: University of California Press, 2007).

[26] Wirzba, "Grace of Good Food," 63.

[27] Scott Sabin, "Environmental Emigration: The World on Our Doorstep," *Creation Care* 37 (Fall 2008): 37-38.

[28] "Scott Sabin," Plant with Purpose, https://plantwithpurpose.org/our_team/scott-sabin/.

[29] "The Economics of Ecosystems and Biodiversity" report (TEEB) was presented in an interim report at a high-level segment of the ninth meeting of the Conference of the

Parties to the Convention on Biological Diversity (CBD COP-9) in Bonn, Germany, in May 2008. Work is ongoing. See the Economics of Ecosystems and Biodiversity website, www.teebweb.org/our-publications/.

[30] Pavan Sukhdev, "The Economics of Ecosystems and Biodiversity—TEEB," EurekAlert!, May 29, 2008, www.eurekalert.org/pub_releases/2008-05/haog-teo052908.php.

[31] Sukhdev, "Economics of Ecosystems and Biodiversity."

[32] Sukhdev, "Economics of Ecosystems and Biodiversity."

[33] As Nathan McClintock explains, when Europeans introduced coffee to Haiti in 1730, forests were cleared and monoculture agriculture began. Clean cultivation between the rows of coffee, indigo, tobacco, and sugarcane depleted the soil and led to erosion (Nathan C. McClintock, "Agroforestry and Sustainable Resource Conservation in Haiti: A Case Study," 2003, http://works.bepress.com/nathan_mcclintock/14/). See as well John Dale Zach Lea, "Charcoal Is Not the Cause of Haiti's Deforestation," *Haïti Liberté*, January 25, 2017, https://haitiliberte.com/charcoal-is-not-the-cause-of-haitis-deforestation/.

[34] "Poverty," GlobalSecurity.org, www.globalsecurity.org/military/world/haiti/poverty.htm.

[35] Mercer Quality of Living Survey of 2016. Mercer utilizes thirty-nine factors in its assessments. These include political, economic, environmental, personal safety, health, education, transportation, and other public-service factors. See Lianna Brinded, "The 29 Cities with the Worst Quality of Life in the World," *Business Insider*, March 1, 2016, www.businessinsider.com/mercers-quality-of-living-index-worst-cities-2016-3.

[36] See the Emmaus University website and mission here: https://emmaus.edu.ht.

[37] Matt Ayers, president of Emmaus University, Cape Haitian, Haiti (personal communication, March 4, 2019). See more at "Capital Facts for Port-au-Prince, Haiti," World's Capital Cities, www.worldscapitalcities.com/capital-facts-for-port-au-prince-haiti/.

[38] Neal Carlstrom, World Venture Missionary—Madagascar (personal communication, March 5, 2019).

[39] Carlstrom, personal communication.

[40] Danielle Carlstrom, World Venture Missionary—Madagascar (personal communication, March 5, 2019).

[41] "What Is Mountaintop Removal Coal Mining?," iLoveMountains.org, http://ilovemountains.org/resources. In other parts of the world the monstrous "bucket wheel excavator" is utilized; the largest on record, from Bogatyr Mine near Ekibastuz, Kazakhstan, weighs in at forty-five thousand tons and has a blade the size of a four-story building (Julian Robinson, "You Have to See This Saw!," *Daily Mail*, January 28, 2015, www.dailymail.co.uk/news/article-2929726/You-saw-Incredible-45-000-ton-machine-4-500-tons-coal-blade-size-four-storey-building.html).

[42] "The Big Muskie was the World's Largest Dragline and one of the seven engineering wonders of the world! The machine has even been featured on the History Channel. The Bucket weighs 460,000 pounds empty and when loaded carried an additional 640,000 pounds. It's volume is equal to that of a 12 car garage. Can you imagine what Big Muskie must have been like to even move such an object, let alone maneuver it effectively?" Noble County Ohio, www.noblecountyohio.com/muskie.html. See as well the Wikipedia article on "Big Muskie," https://en.wikipedia.org/wiki/Big_Muskie.

43 Erik Reece, "Mountaintop-Removal Mining Is Devastating Appalachia, but Residents Are Fighting Back," *Grist*, February 17, 2006, www.grist.org/news/maindish/2006/02/16/reece.htm.

44 M. A. Palmer et al., "Mountaintop Mining Consequences," *Science* 327, no. 5962 (2010): 148-49, https://science.sciencemag.org/content/327/5962/148.

45 James Bruggers, "Mountaintop Mining Is Destroying More Land for Less Coal, Study Finds," *InsideClimate News*, July 26, 2018, https://insideclimatenews.org/news/25072018/appalachia-mountaintop-removal-coal-strip-mining-satellite-maps-environmental-impacts-data.

46 Palmer et al., "Mountaintop Mining Consequences," 148-49, emphasis added.

47 Although not exclusively related to MTR-VF coal mining, see Howard Berkes et al., "An Epidemic Is Killing Thousands of Coal Miners. Regulators Could Have Stopped It," *All Things Considered*, December 18, 2018, www.npr.org/2018/12/18/675253856/an-epidemic-is-killing-thousands-of-coal-miners-regulators-could-have-stopped-it.

48 On June 1, 2011, shareholders of Alpha Natural Resources agreed to buy Massey Energy for $7.1 billion. This came after years of legislation documenting numerous clean-water violations on the part of Massey. Many believed that Massey managers had engineered the sale of the company to protect themselves from liabilities and had arranged new management jobs with Alpha (Clifford Kraus, "Shareholders Approve Massey Energy Sale to Alpha," *New York Times*, June 11, 2011, www.nytimes.com/2011/06/02/business/02coal.html).

49 Alliance for Appalachia, *Mountaintop Removal Facts*, booklet, End Mountaintop Removal Lobby Week, March 2009; "PBS Moyers on America, Is God Green 2006 TVRip SoS," YouTube, uploaded December 31, 2016, www.youtube.com/watch?v=jwMsDVVahTA.

50 This is the same list of toxins identified by Palmer et al. and reported in "Mountaintop Mining Consequences," 148-49.

51 Carol Morello, "Child's Death by Mine Boulder Sets Off Avalanche of Rage," *Chicago Tribune*, January 5, 2005, www.chicagotribune.com/news/ct-xpm-2005-01-09-0501090359-story.html; cf. Tim Thornton, "Family of Boy Killed by Boulder Sues," *Roanoke Times Daily Press*, July 6, 2006, www.dailypress.com/news/dp-xpm-20060706-2006-07-06-0607060311-story.html.

52 Visit the Christians for the Mountains website: www.christiansforthemountains.org/.

53 See a farewell tribute to Larry Gibson and his extraordinary resistance at "Farewell to Larry Gibson, an Appalachian Hero," iLoveMountains.org, September 12, 2012, http://ilovemountains.org/news/3185.

7 THE PEOPLE OF THE NEW COVENANT AND OUR LANDLORD

1 Timothy Larsen, "Defining and Locating Evangelicalism," in *The Cambridge Companion to Evangelical Theology*, ed. Timothy Larsen and Daniel J. Treier (Cambridge: Cambridge University Press, 2007), 5.

2 See Kenneth J. Collins, *The Scripture Way of Salvation: The Heart of John Wesley's Theology* (Nashville: Abingdon, 1997).

3 Larsen, "Defining and Locating Evangelicalism," 1.

4 Riley Case, "In Celebration of Martin Luther and 500 Years of Evangelical Faith," *ENTOS Newsletter*, vol. 23, November 2017.

⁵Douglas Moo ("Nature in the New Creation: New Testament Eschatology and the Environment, *Journal of the Evangelical Theological Society* 49, no. 3 [2006]: 449-58) states that a number of interpreters identify a gulf between Old Testament theology, which embraced nature, and New Testament theology which is "indifferent or even hostile to nature" (453). It is alleged that the New Testament, "under the influence of Greek dualistic notions, has separated humans from their environment," being only "concerned with the salvation of the soul, while 'this world' is viewed quite negatively" (453). As a result it is a "short step from such a matter/spirit dichotomy to the instrumentalist view of nature that is often said to lie at the heart of our environmental crisis" (cf. Ludwig Feuerbach, *The Essence of Christianity* [New York: Harper & Row, 1957]).

⁶See Millard Erickson's classic *Christian Theology* (Grand Rapids: Baker, 1987), 1209-12. Al Truesdale confronts this perspective in "Last Things First: The Impact of Eschatology on Ecology," *Perspectives on Science and Christian Faith* 46 (1994): 116-20.

⁷Lynn White, "The Historical Roots of our Ecological Crisis," *Science* 155 (1967): 1205.

⁸But as Robert Gottlieb states in his introduction to an important anthology on the topic, *This Sacred Earth: Religion, Nature, Environment* (New York: Routledge, 1996), "Religions have been neither simple agents of environmental domination nor unmixed repositories of ecological wisdom. In complex and variable ways, they have been both" (Gottlieb, *This Sacred Earth*, 9). In reality, the abuse of this planet is common to East and West, preindustrial and industrial, Third and First World societies. To quote one author, "It is simpler and surely more accurate to say that human self-seeking is a constant in our natures, and that no culture, no matter what its religion, has managed successfully to eliminate it" (Thomas Sieger Derr, *Environmental Ethics and Christian Humanism*, Abingdon Press Studies in Christian Ethics and Economic Life [Nashville: Abingdon, 1996], 20-21).

⁹See David Kinsley, *Ecology and Religion* (Englewood Cliffs, NJ: Prentice-Hall, 1994), 103-14.

¹⁰In an epoch-changing presentation in 1966 White stated that in the Christian worldview every facet of the physical creation exists only "to serve man's purposes." Thus he credited the modern technological exploitation of nature to the Judeo-Christian ethic (White, "Historical Roots," 1203-7). See Michael Paul Nelson, "The Long Reach of Lynn White Jr.'s 'The Historical Roots of Our Ecologic Crisis,'" *Ecology and Evolution*, December 13, 2016, https://natureecoevocommunity.nature.com/users/24738-michael -paul-nelson/posts/14041-the-long-reach-of-lynn-white-jr-s-the-historical-roots-of -our-ecologic-crisis. White's argument focused on the philosophical and ethical framework created by Christianity, which he believed has led to our current posture of exploitation (See White, "Historical Roots," 1205-6).

¹¹See Sandra L. Richter, "Environmental Law: Wisdom from the Ancients," *Bulletin for Biblical Research* 24, no. 3 (2014): 307-29.

¹²Frederick W. Danker et al., *Greek-English Lexicon of the New Testament and Other Early Christian Literature*, 3rd ed. (Chicago: University of Chicago Press, 2000), s.v. "τέλος," meaning 2 (998).

¹³"The key phrase describing God's approach through the garden, traditionally translated 'in the cool of the day,' should be rendered 'as the Spirit of the day.' 'Spirit' here denotes the theophanic Glory, as it does in Genesis 1:2 and elsewhere in Scripture. And 'the day' has the connotation it often has in the prophets' forecasts of the great coming

judgment (cf. also Judg 11:27 and 1 Cor 4:3). Here in Genesis 3:8 is the original day of the Lord, which served as the prototypal mold in which subsequent pictures of other days of the Lord were cast" (Meredith G. Kline, *Kingdom Prologue: Genesis Foundations for a Covenantal Worldview* [Eugene, OR: Wipf & Stock, 2006], 129).

[14] Gerhard von Rad's classic work is the best resource here: *Old Testament Theology*, Old Testament Library (Louisville, KY: John Knox, 1965), 2:119-28.

[15] Sandra L. Richter, *The Epic of Eden: A Christian Entry into the Old Testament* (Downers Grove, IL: IVP Academic, 2008), 104.

[16] Danker et al., *Greek-English Lexicon of the New Testament*, s.v. "παρουσία," (535).

[17] See George E. Ladd, "Apocalyptic," in *Evangelical Dictionary of Theology* (Grand Rapids: Baker, 2001), 75-79.

[18] Moo, "Nature in the New Creation," 465.

[19] Ben Witherington III, *The Indelible Image: The Theological and Ethical Thought World of the New Testament* (Downers Grove, IL: IVP Academic, 2009), 1:797-805; Richard Bauckham, *Jude, 2 Peter*, Word Biblical Commentary 50 (Waco, TX: Word, 1983), 299; Bauckham, "Jesus and the Wild Animals (Mark 1:13): A Christological Image for an Ecological Age," in *Jesus of Nazareth: Essays on the Historical Jesus and New Testament Christology*, ed. Joel B. Green and Max Turner (Grand Rapids: Eerdmans, 1994), 4; H. Paul Santmire, "Partnership with Nature According to the Scriptures: Beyond the Theology of Stewardship," *Christian Scholars Review* 32, no. 4 (2003): 381-412; John Austin Baker, "Biblical Views of Nature," in *Liberating Life: Contemporary Approaches to Ecological Theology*, ed. Charles Birch, William Eakin, and Jay B. McDaniel (Maryknoll, NY: Orbis, 1990); G. K. Beale, "The Eschatological Concept of New Testament Theology," in *"The Reader Must Understand": Eschatology in Bible and Theology*, ed. K. E. Brower and M. W. Elliott (Leicester, UK: Inter-Varsity Press, 1997), 11-52; James D. G. Dunn, *A Theology of Paul the Apostle* (Grand Rapids: Eerdmans, 1998), 101; John Murray, *The Epistle to the Romans*, New International Commentary on the New Testament (Grand Rapids: Eerdmans, 1968), 303-5.

[20] James D. G. Dunn, *Romans 1–8*, Word Biblical Commentary 38A (Nashville: Thomas Nelson, 1988), xliv.

[21] "The Greek word suggests that creation has been unable to attain the purpose for which it was created" (Moo, "Nature in the New Creation," 461); "The 'vanity' to which creation was subjected would appear to refer to the lack of vitality which inhibits the order of nature and the frustration which the forces of nature meet with in achieving their proper ends" (Murray, *Epistle to the Romans*, 303); "For ages ago Creation was condemned to have its energies marred and frustrated" (William Sanday and Arthur C. Headlam, *The Epistle to the Romans*, 4th ed., International Critical Commentary [Edinburgh: T&T Clark, 1900], 205).

[22] John Stott quotes Günther Bornkamm speaking of the epistle to the Romans as "the last will and testament of the Apostle Paul" (*The Message of Romans: God's Good News for the World* [Leicester, UK: Inter-Varsity Press, 1994], 32).

[23] Moo, "Nature in the New Creation," 462.

[24] Richter, *Epic of Eden*, 119-36.

[25] G. K. Beale, *The Book of Revelation*, New International Greek Testament Commentary (Grand Rapids: Eerdmans, 1999), 1040.

CONCLUSION

[1] Curwood interview with Speth.

[2] Douglas Moo, "Nature in the New Creation: New Testament Eschatology and the Environment," *Journal of the Evangelical Theological Society* 49, no. 3 (2006): 484.

APPENDIX

[1] Eric Holthaus, "Climate Change Blues," *Sierra*, March/April 2018, 60.

BIBLIOGRAPHY

BOOKS AND ARTICLES

Allen, James A. "Reforestation of Bottomland Hardwoods and the Issue of Woody Species Diversity." *Restoration Ecology* 5, no. 2 (June 1997): 125-34.

Allen, Susan. "4 Modern Milking Parlor Designs." Dairy Discovery Zone. December 13, 2017. www.dairydiscoveryzone.com/blog/4-modern-milking-parlor-designs.

Alliance for Appalachia. *Mountaintop Removal Facts*. Booklet. End Mountaintop Removal Lobby Week, March 2009.

American Farmland Trust. "Farms Under Threat." https://www.farmland.org/initiatives/farms-under-threat.

American Society for the Prevention of Cruelty to Animals (ASPCA). "Animals on Factory Farms." www.aspca.org/animal-cruelty/farm-animal-welfare/animals-factory-farms.

———. "A Growing Problem: Selective Breeding in the Chicken Industry; The Case for Slower Growth." www.aspca.org/sites/default/files/chix_white_paper_nov2015_lores.pdf.

———. "What Is Ag-Gag Legislation?" www.aspca.org/animal-protection/public-policy/what-ag-gag-legislation.

Anand, Ash. "Vietnam's Horrific Legacy: The Children of Agent Orange." *News Corp Australia*. May 25, 2015. www.news.com.au/world/asia/vietnams-horrific-legacy-the-children-of-agent-orange/news-story/c008ff36ee3e840b005405a55e21a3e1.

Animal and Plant Health Inspection Service of the USDA. "Overview of U.S. Livestock, Poultry, and Aquaculture Production in 2010 and Statistics on Major Commodities." www.aphis.usda.gov/animal_health/nahms/downloads/Demographics2010_rev.pdf.

Arnold, Bill T. and Bryan E. Beyer, eds. *Readings from the Ancient Near East: Primary Sources for Old Testament Study*. Grand Rapids: Baker Academic, 2002.

Baer, Richard A., Jr. "The Church and Man's Relationship to His Natural Environment." *Quaker Life*, January 1970.

Baker, John Austin. "Biblical Views of Nature." In *Liberating Life: Contemporary Approaches to Ecological Theology*. Edited by Charles Birch, William Eakin, and Jay B. McDaniel, 9-26. Maryknoll, NY: Orbis, 1990.

Barclay, Eliza. "'Piglet Smoothie' Fed to Sows to Prevent Disease; Activists Outraged." NPR. *The Salt*. February 20, 2014. www.npr.org/sections/thesalt/2014/02/20/280183550/piglet-smoothie-fed-to-sows-to-prevent-disease-activists-outraged.

Barringer, Felicity. "Endangered Species Act Faces Broad New Challenges." *New York Times*. June 26, 2005. www.nytimes.com/2005/06/26/politics/endangered-species-act-faces-broad-new-challenges.html.

Bauckham, Richard J. "Jesus and the Wild Animals (Mark 1:13): A Christological Image for an Ecological Age." In *Jesus of Nazareth: Essays on the Historical Jesus and New Testament Christology*. Edited by Joel B. Green and Max Turner, 3-21. Grand Rapids: Eerdmans, 1994.

———. *Jude, 2 Peter*. Word Biblical Commentary 50. Waco, TX: Word, 1983.

Beale, G. K. *The Book of Revelation*. New International Greek Testament Commentary. Grand Rapids: Eerdmans, 1999.

———. "The Eschatological Concept of New Testament Theology." In *"The Reader Must Understand": Eschatology in Bible and Theology*, edited by K. E. Brower and M. W. Elliott, 11-52. Leicester, UK: Inter-Varsity Press, 1997.

Berkes, Howard et al. "An Epidemic Is Killing Thousands of Coal Miners. Regulators Could Have Stopped It." *All Things Considered*. December 18, 2018. www.npr.org/2018/12/18/675253856/an-epidemic-is-killing-thousands-of-coal-miners-regulators-could-have-stopped-it.

Berry, Wendell. *The Unsettling of America: Culture and Agriculture*. 3rd ed. San Francisco: Sierra Club Books, 1996.

Blenkinsopp, Joseph. "The Family in First Temple Israel." In Leo G. Perdue et al., *Families in Ancient Israel*, 48-103. The Family, Religion, and Culture. Louisville, KY: Westminster John Knox, 1997.

Blocher, Henri. *In the Beginning: The Opening Chapters of Genesis*. Downers Grove, IL: InterVarsity Press, 1984.

Block, Daniel I. "All Creatures Great and Small: Recovering a Deuteronomic Theology of Animals." In *The Old Testament in the Life of God's People: Essays in Honor of Elmer A. Martens*, edited by J. Isaak, 200-236. Winona Lake, IN: Eisenbrauns, 2009.

———. "Toward a Biblical Understanding of Humanity's Responsibility in the Face of the Biodiversity Crisis." In *Keeping God's Earth: The Global Environment in Biblical Perspective*, edited by Noah J. Toly and Daniel I. Block, 116-40. Downers Grove, IL: InterVarsity Press, 2010.

Blumenthal, Ralph. "Veterans Accept $180 Million Pact on Agent Orange." *New York Times*. June 16, 2019. www.nytimes.com/1984/05/08/nyregion/veterans-accept-180-million-pact-on-agent-orange.html.

Borowski, Oded. *Agriculture in Iron Age Israel*. Boston: American Schools of Oriental Research, 2002.

———. *Daily Life in Biblical Times*. Archaeology and Biblical Studies 5. Atlanta: Society of Biblical Literature, 2003.

———. *Every Living Thing: The Daily Use of Animals in Ancient Israel*. Lanham, MD: AltaMira, 1999.

Breniquet, Catherine and Cécile Michel, eds. *Wool Economy in the Ancient Near East and the Aegean: From the Beginnings of Sheep Husbandry to Institutional Textile Industry*. Ancient Textile Series 17. Oxford: Oxbow, 2014.

Brinded, Lianna. "The 29 Cities with the Worst Quality of Life in the World." *Business Insider*. March 1, 2016. www.businessinsider.com/mercers-quality-of-living-index-worst-cities-2016-3.

Broom, D. M., M. T. Mendl, and A. J. Zanella. "A Comparison of the Welfare of Sows in Different Housing Conditions." *Animal Science* 61 (1995): 369-85.

Bruggers, James. "Mountaintop Mining Is Destroying More Land for Less Coal, Study Finds." *InsideClimate News*. July 26, 2018. https://insideclimatenews.org/news/25072018/appalachia-mountaintop-removal-coal-strip-mining-satellite-maps-environmental-impacts-data.

Brumm, Mike. "Wean-to-Finish Systems: An Overview." National Hog Farmer. October 1, 1999. www.nationalhogfarmer.com/mag/farming_weantofinish_systems_overview.

Case, Riley. "In Celebration of Martin Luther and 500 Years of Evangelical Faith." *ENTOS Newsletter* vol. 23, November 2017.

Christensen, Duane L. *Deuteronomy 21:10–34:12*. Word Biblical Commentary. Nashville: Thomas Nelson, 2002.

Coblentz, Bonnie A. "Black Bear Numbers Rising in Mississippi." Mississippi State University Extension, July 28, 2005. http://extension.msstate.edu/news/feature-story/2005/black-bear-numbers-rising-mississippi.

Cole, Steven W. "The Destruction of Orchards in Assyrian Warfare." In *Assyria 1995: Proceedings of the 10th Anniversary Symposium of the Neo-Assyrian Text Corpus Project Helsinki, September 7-11, 1995*, edited by S. Parpola and R. M. Whiting. Helsinki: The Neo-Assyrian Text Corpus Project, 1997.

Collins, Kenneth J. *The Scripture Way of Salvation: The Heart of John Wesley's Theology*. Nashville: Abingdon, 1997.

Columbia Water Center. Earth Institute Columbia University. "Punjab, India." http://water.columbia.edu/research-projects/india/punjab-india/.

Dahl, Gudrun and Anders Hjort. *Having Herds: Pastoral Herd Growth and Household Economy*. Stockholm Studies in Social Anthropology 2. Stockholm: Dept. of Social Anthropology, University of Stockholm, 1976.

Danker, Frederick W., et al. *Greek-English Lexicon of the New Testament and Other Early Christian Literature*. 3rd ed. Chicago: University of Chicago Press, 2000.

Derr, Thomas Sieger. *Environmental Ethics and Christian Humanism*. Abingdon Press Studies in Christian Ethics and Economic Life. Nashville: Abingdon, 1996.

Donin, Rabbi Hayim Halevy. *To Be a Jew: A Guide to Jewish Observance in Contemporary Life*. New York: Basic Books, 1972.

Dunn, James D. G. *Romans 1–8*. Word Biblical Commentary 38A. Nashville: Thomas Nelson, 1988.

———. *A Theology of Paul the Apostle*. Grand Rapids: Eerdmans, 1998.

Dwernychuk, Wayne. "Agent Orange and Dioxin Hot Spots in Vietnam." Persistent Organic Pollutants Toolkit. www.popstoolkit.com/about/articles/aodioxinhotspotsvietnam.aspx.

Erickson, Millard. *Christian Theology*. Grand Rapids: Baker, 1987.

Eyre, Christopher. "The Agricultural Cycle, Farming and Water Management in the Ancient Near East." In vol. 1 of *Civilizations of the Ancient Near East*, edited by Jack M. Sasson, 175-89. 1995. Repr., Peabody, MA: Hendrickson, 2006.

Farm Families of Mississippi. "Mississippi's Heritage. Mississippi's Future." https://growingmississippi.org/agriculture-in-mississippi/.

Faust, Avraham. *The Archaeology of Israelite Society in Iron Age II*. Winona Lake, IN: Eisenbrauns, 2012.

———. "Cities, Villages, and Farmsteads: The Landscape of Leviticus 25.29-31." In *Exploring the Longue Durée: Essays in Honor of Lawrence E. Stager*, edited by J. David Schloen, 103-12. Winona Lake, IN: Eisenbrauns, 2009.

Feldmann, Terry. "Equipment, Facility Designs." National Hog Farmer. October 1, 1999. www.nationalhogfarmer.com/mag/farming_equipment_facility_designs.

Feuerbach, Ludwig. *The Essence of Christianity*. New York: Harper & Row, 1957.

Finch, Virginia A., et al. "Why Black Goats in Hot Deserts? Effects of Coat Color on Heat Exchanges of Wild and Domestic Goats." *Physiological Zoology* 53, no. 1 (1980): 19-25.

Food and Agriculture Organization of the United Nations. "Mixed Crop-Livestock Farming: A Review of Traditional Technologies Based on Literature and Field

Experience." FAO Animal Production and Health Papers 152. www.fao.org/3/Y0501E/y0501e00.htm#toc.

Frame, Grant, ed. *Rulers of Babylonia from the Second Dynasty of Isin to the End of Assyrian Domination (1157–612 BC)*. The Royal Inscriptions of Mesopotamia, Babylonian Periods 2. Toronto: University of Toronto Press, 1995.

Gadd, C. J. *The Assyrian Sculptures*. The British Museum. London: Harrison & Sons, 1934.

Gaud, William Gaud. "The Green Revolution: Accomplishments and Apprehensions." AgBioWorld. March 8, 1968. www.agbioworld.org/biotech-info/topics/borlaug/borlaug-green.html.

Glick, Daniel. "Putting the 'Public' Back in Public Lands: An Open Letter to the Next President." *National Wildlife*. October/November, 2008.

GlobalSecurity.org. "Poverty." www.globalsecurity.org/military/world/haiti/poverty.htm.

Goodman, Jim. "Dairy Farming Is Dying. After 40 Years, I'm Done." *Washington Post*. December 21, 2018. www.washingtonpost.com/outlook/dairy-farming-is-dying-after-40-years-im-out/2018/12/21/79cd63e4-0314-11e9-b6a9-0aa5c2fcc9e4_story.html?utm_term=.d4192c32708c.

Gottlieb, Robert. *This Sacred Earth: Religion, Nature, Environment*. New York: Routledge, 1996.

Griffiths, Philip Jones. *Agent Orange: Collateral Damage in Vietnam*. London: Trolley, 2004.

Hahn, Jonathan, "The Tragedy and Wonder of Louisiana's Wetlands: Photographer Ben Depp Uses a Paraglider to Document a Fading Ecosystem." Sierra Club. www.sierraclub.org/sierra/slideshow/tragedy-and-wonder-louisiana-s-wetlands.

Hasel, Michael. *Military Practice and Polemic: Israel's Laws of Warfare in Near Eastern Perspective*. Berrien Springs, MI: Andrews University Press, 2005.

Heschel, Abraham. *The Sabbath: Its Meaning for Modern Man*. New York: Farrar, Straus & Giroux, 1951.

Hesse, Brian. "Animal Husbandry and Human Diet in the Ancient Near East." In vol. 1 of *Civilizations of the Ancient Near East*, edited by Jack M. Sasson, 203-22. 1995. Repr., Peabody, MA: Hendrickson, 2006.

History. "Agent Orange." August 2, 2011. www.history.com/topics/vietnam-war/agent-orange-1.

Holthaus, Eric. "Climate Change Blues." *Sierra*, March/April 2018.

Hopkins, David C. "The Dynamics of Agriculture in Monarchical Israel." In *Society of Biblical Literature 1983 Seminar Papers*, edited by Kent Harold Richards, 177-202. Chico, CA: Scholars Press, 1983.

———. *The Highlands of Canaan: Agricultural Life in the Early Iron Age*. Social World of Biblical Antiquity 3. Decatur, GA: Almond, 1985.

———. "Life on the Land: The Subsistence Struggles of Early Israel." *Biblical Archaeologist* 50 (1987): 178-91.

Humane Society of the United States. "Cage-Free vs. Battery-Cage Eggs." www.humanesociety.org/resources/cage-free-vs-battery-cage-eggs.

———. "An HSUS Report: Welfare Issues with Gestation Crates for Pregnant Sows." February 2013. www.humanesociety.org/sites/default/files/docs/hsus-report-gestation-crates-for-pregnant-sows.pdf.

———. "Shocking Animal Abuse Uncovered at Country's Second Largest Chicken Producer." Press release. June 27, 2017. www.humanesociety.org/news/shocking-animal-abuse-uncovered-countrys-second-largest-chicken-producer.

iLoveMountains.org. "Farewell to Larry Gibson, an Appalachian Hero." September 12, 2012. http://ilovemountains.org/news/3185.

———. "What Is Mountaintop Removal Coal Mining?" http://ilovemountains.org/resources.

India Population 2019. "Population of India 2019, Most Populated States." www.indiapopulation2019.in.

Jacobsen, Thorkild, and Robert M. Adams. "Salt and Silt in Ancient Mesopotamian Agriculture." *Science* 128 (1958): 1251-58.

King, Philip J., and Lawrence E. Stager. *Life in Biblical Israel*. Louisville, KY: Westminster John Knox, 2001.

Kinsley, David. *Ecology and Religion*. Englewood Cliffs, NJ: Prentice-Hall, 1994.

Kirshenbaum, Noam. *Mammals of Israel: A Pocket Guide to Mammals and Their Tracks*. Nature in Israel Series. The National Parks Authority, 2005.

Kline, Meredith G. *Kingdom Prologue: Genesis Foundations for a Covenantal Worldview*. Eugene, OR: Wipf & Stock, 2006.

Koehler, Ludwig, Walter Baumgartner, and Johann J. Stamm. *The Hebrew and Aramaic Lexicon of the Old Testament*. Translated and edited under the supervision of Mervyn E. J. Richardson. 4 vols. Leiden: Brill, 1994-1999.

Kohli, Deepali Singhal, and Nirvikar Singh. "The Green Revolution in Punjab, India: The Economics of Technological Change." Paper presented at Agriculture of the Punjab conference at the Southern Asian Institute, Columbia University. April 1, 1995; revised September 1997. http://people.ucsc.edu/~boxjenk/greenrev.pdf.

Kraus, Clifford. "Shareholders Approve Massey Energy Sale to Alpha." *New York Times*. June 11, 2011. www.nytimes.com/2011/06/02/business/02coal.html.

Ladd, George E. "Apocalyptic." In *Evangelical Dictionary of Theology*. 2nd ed. Edited by Walter A. Elwell. Grand Rapids: Baker, 2001.

Ladizinsky, Gideon. "Origin and Domestication of the Southwest Asian Grain Legumes." In *Foraging and Farming: The Evolution of Plant Exploitation*, edited by David R. Harris and Gordon C. Hillman, 374-89. London: Unwin Hyman, 1989.

Laniak, Timothy S. *Shepherds After My Own Heart: Pastoral Traditions and Leadership in the Bible*, New Studies in Biblical Theology 20. Downers Grove, IL: InterVarsity Press, 2006.

———. *While Shepherds Watch Their Flocks: Forty Daily Reflections on Biblical Leadership*. n.p.: ShepherdLeader Publications, 2007.

Larsen, Timothy. "Defining and Locating Evangelicalism." In *The Cambridge Companion to Evangelical Theology*, edited by Timothy Larsen and Daniel J. Treier, 1-14. Cambridge: Cambridge University Press, 2007.

Lea, John Dale Zach "Charcoal Is Not the Cause of Haiti's Deforestation." *Haïti Liberté*. January 25, 2017. https://haitiliberte.com/charcoal-is-not-the-cause-of-haitis-deforestation/.

Lee, Roberta. "Summer Fun, but Not for Pigs: The Horror of Gestation Crates and Life in a Factory Farm." *Huffington Post*. July 16, 2015. www.huffpost.com/entry/summer-fun-but-not-for-pi_b_7759466.

Li, Yanxia, Zhonghong Wu, Glen Broderick, and Brian Holmes. "Rapid Assessment of Feed and Manure Nutrient Management on Confinement Dairy Farms." *Nutrient Cycling in Agroecosystems* 82, no. 2 (2008): 107-15.

Limaye, Yogita. "Will Organic Revolution Boost Farming in India?" *BBC News*. September 24, 2018. www.bbc.com/news/av/business-45605018/will-organic-revolution-boost-farming-in-india.

Lovaas, Deron. "Measuring Suburban Sprawl." National Resource Defense Council. April 2, 2014. https://www.nrdc.org/experts/deron-lovaas/measuring-suburban-sprawl.

Luckenbill, Daniel David, ed. *Ancient Records of Assyria and Babylonia*. Chicago: University of Chicago Press, 1926–1927. Repr., New York: Greenwood, 1968.

MacDonald, Nathan. *What Did the Ancient Israelites Eat? Diet in Biblical Times*. Grand Rapids: Eerdmans, 2008.

Maeir, Aren M., Oren Ackermann, and Hendrik J. Bruins. "The Ecological Consequences of a Siege: A Marginal Note on Deuteronomy 20:19-20." In *Confronting the Past: Archaeological and Historical Essays on Ancient Israel in Honor of William G. Dever*. Edited by Seymour Gitin, J. Edward Wright, and J. P Dessel, 239-43. Winona Lake, IN: Eisenbrauns, 2006.

Martini, Edwin. *Agent Orange: History, Science, and the Politics of Uncertainty*. Boston: University of Massachusetts Press, 2012.

McBride, S. Dean. "Polity of the Covenant People: The Book of Deuteronomy." In *A Song of Power and the Power of Song: Essays on the Book of Deuteronomy*, edited by Duane L. Christensen, 62-77. Winona Lake, IN: Eisenbrauns, 1993.

McClintock, Nathan C. "Agroforestry and Sustainable Resource Conservation in Haiti: A Case Study." 2003. http://works.bepress.com/nathan_mcclintock/14/.

McConville, J. Gordon. *Deuteronomy*. Apollos Old Testament Commentaries 5. Leicester, UK: Inter-Varsity Press, 2002.

McDowell, Catherine. *The Image of God in the Garden of Eden: The Creation of Humankind in Genesis 2:5–3:24 in Light of the* mīs pî, pīt pî, *and* wpt-r *Rituals of Mesopotamia and Ancient Egypt*. Siphrut 15. Winona Lake, IN: Eisenbrauns, 2015.

McKenna, Maryn. "By 2020, Male Chicks May Avoid Death by Grinder." *National Geographic*. June 13, 2016. www.nationalgeographic.com/people-and-culture/food/the-plate/2016/06/by-2020--male-chicks-could-avoid-death-by-grinder/.

Meyers, Carol. "The Family in Early Israel." In *Families in Ancient Israel*, edited by Leo G. Perdue et al., 1-47. The Family, Religion, and Culture. Louisville, KY: Westminster John Knox, 1997.

Milgrom, Jacob. *Leviticus: A Book of Ritual and Ethics*. Continental Commentaries. Minneapolis: Fortress, 2004.

Military Wikia. "Agent Orange." https://military.wikia.org/wiki/Agent_Orange.

Miller, Patrick D. *Chieftains of the Highland Clans: A History of Israel in the 12th and 11th Centuries B.C.* Grand Rapids: Eerdmans, 2005.

———. "Complex Chiefdom Model." In *Chieftains of the Highland Clans: A History of Israel in the 12th and 11th Centuries B.C*, 6-28. Grand Rapids: Eerdmans, 2005.

———. *The Religion of Ancient Israel*. Louisville, KY: Westminster John Knox, 2000.

Mississippi Department of Agriculture and Commerce. "Mississippi Agriculture Overview." www.mdac.ms.gov/agency-info/mississippi-agriculture-snapshot/.

Mississippi Wildlife, Fisheries, and Parks. "Range, Movements and Sightings." www.mdwfp.com/wildlife-hunting/black-bear-program/mississippi-black-bear-ecology/range-movements-and-sightings/.

Montgomery, David R. *Dirt: The Erosion of Civilizations.* Berkeley: University of California Press, 2007.

Moo, Douglas J. "Nature in the New Creation: New Testament Eschatology and the Environment." *Journal of the Evangelical Theological Society* 49, no. 3 (2006): 449-88.

———, and Jonathan A. Moo. *Creation Care: A Biblical Theology of the Natural World.* Grand Rapids: Zondervan, 2018.

Morello, Carol. "Child's Death by Mine Boulder Sets Off Avalanche of Rage." *Chicago Tribune.* January 5, 2005. www.chicagotribune.com/news/ct-xpm-2005-01-09-0501090359-story.html.

Muir, Patricia S. "Consequences for Organic Matter in Soils." http://people.oregonstate.edu/~muirp/orgmater.htm.

Murray, John. *The Epistle to the Romans.* New International Commentary on the New Testament. Grand Rapids: Eerdmans, 1968.

Na, Danny, and Tom Polansek. "U.S. Hogs Fed Pig Remains, Manure to Fend Off Deadly Virus Return." *Scientific American.* www.scientificamerican.com/article/u-s-hogs-fed-pig-remains-manure-to-fend-off-deadly-virus-return/.

Nam, Roger S. *Portrayals of Economic Exchange in the Book of Kings.* Biblical Interpretation Series 112. Leiden: Brill, 2012.

National Agricultural Statistics Service. *Quarterly Hogs and Pigs.* July 27, 2019. www.nass.usda.gov/Publications/Todays_Reports/reports/hgpg0619.pdf.

———. "United States Hog Inventory Up 2 Percent." Accessed June 16, 2019. www.nass.usda.gov/Newsroom/2018/12-20-2018.php.

National Park Service. "The Story of the Teddy Bear." Last updated February 16, 2019. www.nps.gov/thrb/learn/historyculture/storyofteddybear.htm.

Nelson, Michael Paul. "The Long Reach of Lynn White Jr.'s 'The Historical Roots of Our Ecologic Crisis.'" *Ecology and Evolution.* December 13, 2016. https://natureecoevocommunity.nature.com/users/24738-michael-paul-nelson/posts/14041-the-long-reach-of-lynn-white-jr-s-the-historical-roots-of-our-ecologic-crisis.

Nelson, Richard. *Deuteronomy.* Old Testament Library. Louisville, KY: Westminster John Knox, 2002.

Nicoletti, Kimberly. "The Aftermath of Agent Orange: Local Woman Forms Nonprofit to Aid Affected Children." Vietnam Agent Orange Relief and Responsibility Campaign. January 14, 2006. www.vn-agentorange.org/aspen_20060114.html.

North Carolina Farm Families. "The Facts." https://ncfarmfamilies.com/thefacts/.

Oklahoma State University. Department of Animal Science. "Breeds of Livestock—Anatolian Black Goats." www.ansi.okstate.edu/breeds/goats/anatolianblack.

———. Department of Animal Science. "Breeds of Livestock—Awassi Sheep." www.ansi.okstate.edu/breeds/sheep/awassi.

Otzen, Benedikt. "Israel Under the Assyrians." In *Power and Propaganda, Mesopotamia* 7, edited by M. T. Larsen, 251-62. Copenhagen: Akademisk Forlag, 1979.

Palmer, M. A., et al. "Mountaintop Mining Consequences." *Science* 327, no. 5962 (2010): 148-49. https://science.sciencemag.org/content/327/5962/148.

Paul, Shalom M. and William G. Dever. *Biblical Archaeology*. New York: Quadrangle & New York Times, 1974.

"PBS Moyers on America, Is God Green 2006 TVRip SoS." YouTube. Uploaded December 31, 2016 by Andre Emile. www.youtube.com/watch?v=jwMsDVVahTA.

Pepper, Daniel. "The Toxic Consequences of the Green Revolution." *U.S. News and World Report*. July 7, 2008. www.usnews.com/news/world/articles/2008/07/07/the-toxic-consequences-of-the-green-revolution.

Plant with Purpose. "Scott Sabin." https://plantwithpurpose.org/our_team/scott-sabin.

Pollan, Michael. "An Open Letter to the Next Farmer in Chief." *New York Times Magazine*, October 12, 2008. www.nytimes.com/2008/10/12/magazine/12policy-t.html.

Polycarpou, Lakis. "'Small Is Also Beautiful'—Appropriate Technology Cuts Rice Farmers' Water Use by 30 Percent in Punjab India." *State of the Planet* (blog). Columbia University Earth Institute. November 17, 2010. https://blogs.ei.columbia.edu/2010/11/17/"small-is-also-beautiful"---appropriate-technology-cuts-rice-famers'-water-use-by-30-percent-in-punjab-india/.

Powell, Marvin A. "Salt, Seed, and Yields in Sumerian Agriculture: A Critique of the Theory of Progressive Salinization." *Zeitschrift für Assyriologie und Vorderasiatische Archäologie* 75 (1985): 7-38.

Pritchard, James B., ed. *Ancient Near Eastern Texts Relating to the Old Testament*. 3rd ed. Princeton: Princeton University Press, 1969.

Reece, Erik. "Mountaintop-Removal Mining Is Devastating Appalachia, but Residents Are Fighting Back." *Grist*, February 17, 2006. www.grist.org/news/maindish/2006/02/16/reece.htm.

Renfrew, Jane. "Vegetables in the Ancient Near Eastern Diet." In *Civilizations of the Ancient Near East*, edited by Jack M. Sasson, 191-202. 1995. Repr., Peabody, MA: Hendrickson, 2006.

Richter, Sandra L. "The Archaeology of Mt. Ebal and Mt. Gerizim and Why It Matters." In *Sepher Torath Mosheh: Studies in the Composition and Interpretation of Deuteronomy*, edited by Daniel I. Block and Richard L. Schultz, 311-37. Peabody, MA: Hendrickson, 2017.

———. "The Bible and American Environmental Practice: An Ancient Code Addresses a Current Crisis." In *The Bible and the American Future*, edited by Robert Jewett with Wayne L. Alloway Jr. and John G. Lacey, 108-29. Eugene, OR: Cascade, 2009.

———. "A Biblical Theology of Creation Care: Is Environmentalism a Christian Value?" *The Asbury Journal* 62, no. 1 (2007): 67-76.

———. *The Deuteronomistic History and the Name Theology: l^ešakkēn š^emô šām in the Bible and the Ancient Near East*. Beihefte zur Zeitschrift für die alttestamentliche Wissenschaft 318. Berlin: de Gruyter, 2002.

———. "Eighth-Century Issues: The World of Jeroboam II, the Fall of Samaria, and the Reign of Hezekiah." In *Ancient Israel's History: An Introduction to Issues and Sources*, edited by Bill T. Arnold and Richard S. Hess. Grand Rapids: Baker Academic, 2014.

———. "Environmental Law in Deuteronomy: One Lens on a Biblical Theology of Creation Care." *Bulletin for Biblical Research* 20, no. 3 (2010): 331-54.

———. "Environmental Law: Wisdom from the Ancients." *Bulletin for Biblical Research* 24, no. 3 (2014): 307-29.

———. "Environmentalism and the Evangelical: Just the Bible for Those Justly Concerned." *Westmont Magazine*, Spring 2019.

———. *The Epic of Eden: A Christian Entry into the Old Testament*. Downers Grove, IL: IVP Academic, 2008.

———. *The Epic of Eden: Isaiah*. One Book. Franklin, TN: Seedbed, 2016.

———. "The Place of the Name in Deuteronomy." *Vetus Testamentum* 57 (2007): 342-66.

———. "The Question of Provenance and the Economics of Deuteronomy." *Journal for the Study of the Old Testament* 42 (2017): 23-50.

———. "Religion and the Environment." In *Handbook of Religion: A Christian Engagement with Traditions, Teachings, and Practices*, edited by Terry C. Muck, Harold A. Netland, and Gerald R. McDermott, 746-55. Grand Rapids: Baker Academic, 2014.

———. "The Servant and the Idol." In *The Epic of Eden: Isaiah*, 119-38. One Book. Franklin, TN: Seedbed, 2016.

Robinson, Julian. "You Have to See this Saw!" *Daily Mail*. January 28, 2015. www.dailymail.co.uk/news/article-2929726/You-saw-Incredible-45-000-ton-machine-4-500-tons-coal-blade-size-four-storey-building.html.

Rosen, Baruch. "Subsistence Economy in Iron Age I." In *From Nomadism to Monarchy: Archaeological and Historical Aspects of Early Israel*, edited by Israel Finkelstein and Nadav Na'aman, 339-51. Jerusalem: Israel Exploration Society, 1994.

———. "Subsistence Economy of Stratum II." In *'Izbet Ṣarṭah: An Early Iron Age Site Near Rosh Ha'ayin, Israel*. Edited by Israel Finkelstein, 156-85. BAR International Series 299. Oxford: B.A.R., 1986.

Runyon, Luke. "Judge Strikes Down Idaho 'Ag-Gag' Law, Raising Questions for Other States." NPR. August 4, 2015. www.npr.org/sections/thesalt/2015/08/04/429345939/idaho-strikes-down-ag-gag-law-raising-questions-for-other-states.

Sabin, Scott. "Environmental Emigration: The World on Our Doorstep." *Creation Care* 37 (Fall 2008): 37-38.

Sahlins, Marshall D. *Tribesmen*. Englewood Cliffs, NJ: Prentice Hall, 1968.

Sanday, William, and Arthur C. Headlam. *The Epistle to the Romans*. 4th ed. International Critical Commentary. Edinburgh: T&T Clark, 1900.

Santmire, H. Paul. "Partnership with Nature According to the Scriptures: Beyond the Theology of Stewardship." *Christian Scholars Review* 32, no. 4 (2003): 381-412.

Sasson, Aharon. *Animal Husbandry in Ancient Israel: A Zooarchaeological Perspective on Livestock Exploitation, Herd Management and Economic Strategies*. London: Equinox, 2010.

Sasson, Jack. "Should Cheeseburgers Be Kosher?" *Biblical Research* 19 (2008): 40-43, 50-51.

Schlosser, Eric. "Cheap Food Nation," *Sierra*. November/December 2006. https://vault.sierraclub.org/sierra/200611/cheapfood.asp.

———. "Inside the Slaughterhouse." *Frontline*. www.pbs.org/wgbh/pages/frontline/shows/meat/slaughter/slaughterhouse.html.

Scully, Matthew. *Dominion: The Power of Man, the Suffering of Animals, and the Call to Mercy*. New York: St. Martin's Griffin, 2002.

Simek, Stephanie L., et al. "History and Status of the American Black Bear in Mississippi." *Ursus* 23, no. 2 (2012): 159-67.

Smith, Timothy Lawrence. *Revivalism and Social Reform: American Protestantism on the Eve of the Civil War.* New York: Harper & Row, 1965.

Smith, Virginia. *Clean: A History of Personal Hygiene and Purity.* New York: Oxford University Press, 2007.

Smoak, Jeremy. "Building Houses and Planting Vineyards: The Early Inner-Biblical Discourse on an Ancient Israelite Wartime Curse." *Journal of Biblical Literature* 127, no. 1 (2008): 19-35.

Soede, Nicoline M., and Bas Kemp. "Housing Systems in Pig Husbandry Aimed at Welfare; Consequences for Fertility." An abbreviated and updated version of "Reproductive Issues in Welfare Friendly Housing Systems in Pig Husbandry: A Review." *Reproduction in Domestic Animals* 47 Supplement 5 (2012): 51-57.

Speth, Gus. Interviewed by Steve Curwood. WineWaterWatch.org, February 13, 2015. http://winewaterwatch.org/2016/05/we-scientists-dont-know-how-to-do-that-what-a-commentary.

Stager, Lawrence E. "Archaeology of the Family in Ancient Israel." *Bulletin of the American Schools of Oriental Research* 260 (1985): 1-35.

Starrs, Tom. "Fossil Food: Consuming Our Future." Center for Ecoliteracy. June 29, 2009. www.ecoliteracy.org/article/fossil-food-consuming-our-future.

Stocking, Ben. "Agent Orange Still Haunts Vietnam, US." *Washington Post*, June 14, 2007. www.washingtonpost.com/wp-dyn/content/article/2007/06/14/AR2007061401077.html?noredirect=on.

Stone, Lawson G. "Worship as Cherishing YHWH's World in Leviticus." Paper presented at the Annual Meeting of the Institute for Biblical Research. New Orleans, LA, November 21, 2009.

Stott, John. *The Message of Romans: God's Good News for the World.* Leicester, UK: InterVarsity Press, 1994.

Sukhdev, Pavan. "The Economics of Ecosystems and Biodiversity—TEEB." Interim report, May 29, 2008. Overview at EurekAlert! May 29, 2008. www.eurekalert.org/pub_releases/2008-05/haog-teo052908.php.

Swartz, Daniel. "Jews, Jewish Texts, and Nature: A Brief History." In *This Sacred Earth: Religion, Nature, Environment*, edited by Roger S. Gottlieb, 92-109. New York: Routledge, 1996.

Theocharous, Myrto. "Becoming a Refuge: Sex Trafficking and the People of God." *Journal of the Evangelical Theological Society* 59 (2016): 309-22.

Thornton, Tim. "Family of Boy Killed by Boulder Sues." *Roanoke Times Daily Press.* July 6, 2006. www.dailypress.com/news/dp-xpm-20060706-2006-07-06-0607060311-story.html.

Tigay, Jeffrey. *Deuteronomy.* JPS Torah Commentary. Philadelphia: The Jewish Publication Society, 1996.

Tillion, J. P., and F. Madec. "Diseases Affecting Confined Sows: Data from Epidemiological Observations." *Annales de Recherches Vétérinaires, INRA Editions* 15, no. 2 (1984): 195-99.

Truesdale, Al. "Last Things First: The Impact of Eschatology on Ecology." *Perspectives on Science and Christian Faith* 46 (1994): 116-20.

Tyson Foods. "Contract Poultry Farming." www.tysonfoods.com/who-we-are/our-partners/farmers/contract-poultry-farming.

United States Armed Forces Medical Intelligence Center and the United States Defense Intelligence Agency Deputy Directorate for Scientific and Technical Intelligence. *Venomous Snakes of the Middle East Identification Guide*. Department of Defense Intelligence Document DST-1810S-469-91. Fort Detrick, Frederick, MD: Armed Forces Medical Intelligence Center, 1991.

United States Department of Veterans Affairs. "Veterans' Diseases Associated with Agent Orange." www.publichealth.va.gov/exposures/agentorange/diseases.asp.

United States Department of Veteran's Affairs Office of Public and Intergovernmental Affairs. "Over $2.2 Billion in Retroactive Agent Orange Benefits Paid to 89,000 Vietnam Veterans and Survivors for Presumptive Conditions." August 31, 2011. www.va.gov/opa/pressrel/pressrelease.cfm?id=2154.

University of Florida School of Forest Resources and Conservation. "Bottomland Hardwoods." www.sfrc.ufl.edu/extension/4h/ecosystems/bottomland_hardwoods/bottomland_hardwoods_description.pdf.

Ussishkin, David. "On the So-Called Aramaean 'Siege Trench' in Tell eṣ-Ṣafi, Ancient Gath." *Israel Exploration Journal* 59, no. 2 (2009): 137-57.

Van Houtan, Kyle S. "Extinction and Its Causes." *Creation Care*, Fall 2008.

von Rad, Gerhard. *Old Testament Theology*. Old Testament Library. Louisville, KY: John Knox, 1965.

Ward, Geoffrey C., and Ken Burns. *The Vietnam War: An Intimate History*. New York: Knopf, 2017.

Waskow, Arthur. "What is Eco-Kosher?" In *This Sacred Earth*, edited by Roger S. Gottlieb, 297-300. New York: Routledge, 1996.

Weber, Max. "Bureaucracy." In *Economy and Society: An Outline of Interpretive Sociology*, edited by Guenther Roth and Claus Wittich, 2:956-1005. Berkeley: University of California Press, 1978.

———. "Patriarchalism and Patrimonialism," In *Economy and Society: An Outline of Interpretive Sociology*, edited by Guenther Roth and Claus Wittich, 2:1006-69. Berkeley: University of California Press, 1978.

Webster, John. *Animal Welfare: Limping Towards Eden; A Practical Approach to Redressing the Problem of Our Dominion Over the Animals*. Oxford: Blackwell, 2005.

Weinfeld, Moshe. "בְּרִית b'rît." In *Theological Dictionary of the Old Testament*. Edited by G. Johannes Botterweck and Helmer Ringgren. Translated by John T. Willis et al. Grand Rapids: Eerdmans, 1975.

White, Lynn. "The Historical Roots of our Ecological Crisis." *Science* 155 (1967): 1203-7.

Wikipedia. "Agent Orange." http://en.wikipedia.org/wiki/Agent_Orange.

Wilcox, Fred A. "Toxic Agents: Agent Orange Exposure." In *The Oxford Companion to American Military History*, edited by John Whiteclay Chambers. Oxford: Oxford University Press, 1999.

Winner, Lauren F. *The Mudhouse Sabbath*. Brewster, MA: Paraclete Press, 2003.

Wirzba, Norman. "The Grace of Good Food and the Call to Good Farming." *Review and Expositor* 108 (Winter 2011): 61-71.

———. *The Paradise of God: Renewing Religion in an Ecological Age*. New York: Oxford University Press, 2003.

Witherington, Ben, III. *The Indelible Image: The Theological and Ethical Thought World of the New Testament*. Downers Grove, IL: IVP Academic, 2009.

Woodhouse, Leighton Akio. "Charged with the Crime of Filming a Slaughterhouse." *The Nation*. July 31, 2013. www.thenation.com/article/charged-crime-filming-slaughter house/.

World's Capital Cities. "Capital Facts for Port-au-Prince, Haiti." www.worldscapitalcities.com/capital-facts-for-port-au-prince-haiti/.

Wright, Christopher J. H. *God's People in God's Land: Family, Land, and Property in the Old Testament*. Grand Rapids: Eerdmans, 1990.

Wright, Jacob. Review of *Military Practice and Polemic: Israel's Laws of Warfare in Near Eastern Perspective*, by Michael Hasel. *Journal of Biblical Literature* 125, no. 3 (2006): 577.

———. "Warfare and Wanton Destruction," *Journal of Biblical Literature* 127, no. 3 (2008): 423-58.

Yoffee, Norman. "The Collapse of Ancient Mesopotamian States and Civilization." In *The Collapse of Ancient States and Civilizations*, edited by Norman Yoffee and George L. Cowgill, 44-68. Tucson: University of Arizona Press, 1991.

Young, Brad. "Black Bears in Mississippi Past and Present." *Wildlife Issues* (Fall/Winter 2004), www.mdwfp.com/media/3306/ms_black_bear_wildlife_issues_2004.pdf.

Zwerdling, Daniel. "India's Farming 'Revolution' Heading for Collapse." *All Things Considered*. April 13, 2009. www.npr.org/templates/story/story.php?storyId=102893816.

WEBSITES

Blessed Earth: www.blessedearth.org.

Christians for the Mountains: www.christiansforthemountains.org.

The Economics of Ecosystems and Biodiversity (TEEB) Initiative: http://ec.europa.eu/environment/nature/biodiversity/economics.

The Economics of Ecosystems and Biodiversity (TEEB) Publications: www.teebweb.org/our-publications.

Emmaus University: https://emmaus.edu.ht.

Farm Sanctuary: www.farmsanctuary.org/learn/factory-farming/pigs-used-for-pork.

Noble County Ohio: www.noblecountyohio.com/muskie.html.

People for the Ethical Treatment of Animals (PETA): www.peta.org/students/missions/male-chicks-ground-up-alive.

Plant with Purpose: https://plantwithpurpose.org/board.

Seedbed: www.seedbed.com.

Union Rescue Mission: https://urm.org/solution.

Wikipedia: https://en.wikipedia.org/wiki/Big_Muskie.

ART CREDITS

Figure 1. Seven days of creation in Genesis 1–2:3, originally published in Sandra Richter, *Epic of Eden: A Christian Entry into the Old Testament* (Downers Grove, IL: IVP Academic, 2008), 100. Used with permission from InterVarsity Press.

Figure 2. Map of India, copyright InterVarsity Press.

Figure 3. Oxen threshing wheat, used by permission, courtesy of Reconstruction: Semitic Museum, Harvard University and Estate of L. E. Stager; illustration: C. S. Alexander.

Figure 4. Chickens in industrial coop, photo: ITamar K. / Wikimedia Commons.

Figure 5. Assyrian king kills a lion, photo: Lawson G. Stone / Wikimedia Commons.

Figure 6. Assyrian eunuchs return with the prizes of the royal hunt, photo: Lawson G. Stone / Wikimedia Commons.

Figure 7. Assyrian troops cutting palms, original drawing IV, 41, in Paterson, *Palace of Sinacherib* (1915), pl. 13; cf. Steven W. Cole, "The Destruction of Orchards in Assyrian Warfare," fig. 8, 40.

Figure 8. South Vietnam map, by InterVarsity Press, based on GVN Herbicide, illustration: 718 Bot, English Wikipedia / Wikimedia Commons.

Figure 9. Israelite tribal society, originally published in Sandra Richter, *Epic of Eden: A Christian Entry into the Old Testament* (Downers Grove, IL: IVP Academic, 2008), 26. Used with permission from InterVarsity Press.

Figure 10. Dragline excavator, photo courtesy of Vivian Stockman.

Figure 11. Lunar landscapes, photo courtesy of Vivian Stockman.

SUBJECT INDEX

Agent Orange, 64, 64 fig. 8, 65
ag-gag laws, 42, 43
analogia, 53, 61
apocalyptic, 99, 100
Appalachia, Appalachian, 3, 83, 85, 86, 87
Appalachian Voices, 83
Assyria, Assyrians, 53, 56, 62, 63
Awassi, Awassi sheep, 31, 32
battery cages, 37, 38 fig. 4
battery, hen 37
bêt 'āb, 69, 70, 70 fig. 9, 72, 73, 75, 76. See also father's house(hold)
black bear, 57, 58, 132n23, 132n25
black Sinai goats, 31, 32
bovine(s), 33, 41, 121n5. See also cattle
British Museum, 54, 55, 56, 131n19
broilers, 27, 37, 38
cattle, 20, 31, 33, 41, 42, 45, 126n13. See also bovine(s)
chicken(s), 37, 38, 39, 40, 43, 47. See also poultry
Christians for the Mountains, 85, 88
coal impoundment, 85. See also mountaintop removal coal mining (MTR)
coal mining, 81, 82. See also mountaintop removal coal mining (MTR)
consumerism, 45, 46
covenant, 9, 16, 17, 30, 31, 47, 50, 68, 91, 93, 98, 100, 124n1
day of the Lord [Yahweh], 93, 94, 95, 98, 99, 139-40n13
Delta, 27, 28, 57, 58, 132n25
dominion, 40, 119n5
dragline, 82, 82 fig. 10, 137n4

dry farming, 52, 74, 76, 108
egg(s), 27, 36, 37, 38 fig. 4, 42, 47, 53, 56, 115
Elimelech, 74, 76
Emmaus University, 79, 80
environmental
  abuse, 79
  concern, 2, 7, 14, 107
  crisis, 24, 93, 110, 139n5
  damage, 63
  degradation, 3, 78, 79, 80, 81, 90, 110
  disaster, 86
  domination, 139n8
  impact, 108, 133-34n11
  problems, 106
  stewardship, 1, 3, 5, 7, 15, 92, 105, 106, 111, 114
  terrorism, 60, 62, 63, 65, 66, 108
European Union, 37, 127n22
factory farming; factory farms, 35, 43
fallow, fallowing, 21, 22, 23, 24, 78, 108, 122n13, 122n17, 123n22, 123n25
father's house(hold), 69, 70, 72, 75. See also *bêt 'āb*
Fleece, 18, 32, 71
forced draft urbanization, 64
fossil fuel, 27, 28, 116, 124n36, 128n34
Gath (Tel es-Sâfi), 63, 133n11
gestation crates, 36, 127n20, 127n22
Gnosticism, 92
Go Vap orphanage, 65
Great Awakening, the, 91
Green Revolution, the, 24, 25, 26, 28
Haiti, 78, 79, 80, 81, 111, 137n33
Humane Society, 37, 114, 115, 128n30

hungry season, 19, 34, 76
industrial/industrialized agriculture, 3, 24, 25, 27, 28, 58
John Wesley, 91, 92
land grant, 9, 11, 15, 16, 17, 30, 107, 120n8
levirate law, 72
lion(s), 51, 52, 54, 55, 55 fig. 5, 56, 131n19
Lynn White, 93, 139n10
Madagascar, 3, 80, 81, 111. See also Red Island
mass-confinement animal husbandry, 35, 40, 43, 47, 128n34
milk, 21, 31, 32, 37, 44, 45, 47, 121n5, 131n14
Mississippi, 27, 28, 57, 58, 124n35, 132-33n25
  river, 86, 132n22
monocrop agriculture, 27
mountaintop removal coal mining (MTR), 3, 81, 84, 85, 87, 88, 90, 137n41. See also coal impoundment, coal mining
muzzle the ox, 35, 126n13
Naomi, 74, 75
Neo-Assyrian Empire, 53, 62
nepeš, 41
new earth, 68, 94, 103
No Man's Land, 60
Norman Borlaug, 25
Operation Ranch Hand, 60, 63, 66
orphan(s), 3, 5, 45, 67, 68, 70, 71, 72, 74, 75, 76, 77, 79, 80, 81, 89, 90, 104, 105, 107, 110, 111
Parousia, 98
pars pro toto, 53, 61
patriarch, 69, 76
patriarchal 68, 69, 74

Patricia Muir, 26, 122n16,
123n22-23
pig(s), 36, 126n16,
126-27n17, 127n20
piglets, 36, 37
Plant with Purpose, 78, 117,
122-23n19
poultry, 27, 37, 39, 41,
128-29n40. *See also*
chicken(s)
Punjab (India), 14, 24, 25, 25
fig. 2, 26, 28
Rawl, West Virginia, 84, 85
Red Island, 80. *See also*
Madagascar
resident alien, 68, 70, 76, 77,
89, 90
royal hunt, 56, 56 fig. 6,
131n19
royal hunting park 55

rule, 7, 8, 8 fig. 1, 9, 11, 12, 53,
54, 55, 56, 93, 96, 119n2,
119n5,
Ruth 74, 75, 76
Sabbath, 7, 10, 21, 22, 23, 30,
31, 35, 108, 119n5, 123n25
Sargon II, 62
Shalmaneser III, 63
Siege, 53, 61, 62, 63,
133-34n11
slaughter, 19, 36, 38, 39, 40,
41, 42, 47, 55, 97, 108,
126n15, 127n20, 131n19
slaughterhouse(s), 39, 43,
129n43
subsistence
economy, 17, 33, 52
farmer/farming, 76, 78,
79
level, 135n8

struggles, 77
suzerain, 9, 11, 16
TEEB (The Economics of
Ecosystems and
Biodiversity), 79, 136n29
Tel es-Sâfi. *See* Gath
telos, 93, 96, 97, 111
tribal
culture, 68-69, 72, 74,
135n2, 135n9
society, 68, 70 fig. 9
urban sprawl, 51, 132n22
valley fill(s) (VF), 82, 83
vassal(s), 9, 11, 16, 53, 54
veteran(s), 65, 134n21
Vietnam, 60, 63, 64, 64 fig. 8,
65, 66, 134n13
widow(s), 3, 45, 65, 67, 68, 70,
71, 72, 73, 74, 75, 76, 77, 79,
80, 81, 89, 90, 104, 105, 107

SCRIPTURE INDEX

OLD TESTAMENT

Genesis
1, *9, 12, 97, 119*
1:2, *139*
1:14-19, *8*
1:20-23, *8*
1:24-25, *8*
1:26, *9*
1:28, *9, 11*
2:1-3, *10*
2:5–3:24, *119*
2:15, *11, 109, 112*
3:8, *95, 140*
3:15, *95*
3:16, *12*
3:17-19, *12*
3:19, *12*
9:10-11, *50*
24, *70*
31, *70*
38, *73*

Exodus
20:8, *122*
22:28, *121*
23:10-12, *21*

Leviticus
17, *41*
17:4, *128*
19:9, *76*
19:10, *77*
19:19, *123*
23:22, *76, 77*
25, *131, 135*
25:4-7, *22*
25:13-17, *121*
25:17, *24*
25:23, *24, 121*
25:43, *119*
25:46, *119*
25:53, *119*
26:34-35, *123*
26:43, *123*

Numbers
24:19, *119*
32:22, *119*
32:29, *119*

Deuteronomy
1:8, *16*
4:40, *16*
4:41, *121*
5:12, *122*
5:14-15, *31*
5:33, *24*
7:13, *131*
10:9, *121*
11:14, *131*
12:10-12, *20*
12:12, *121*
12:15, *35, 41, 51*
12:17, *131*
12:22, *35, 41, 51*
14:5, *35, 41, 51*
14:21, *37*
14:22-23, *17*
14:23, *131*
14:27, *121*
14:28-29, *18*
14:29, *121*
15:19, *19*
15:19-20, *17*
15:22, *35, 41, 51*
16:9, *131*
18:1, *121*
18:3-5, *18*
18:4, *131*
19:2, *121*
19:7, *121*
20:19, *60*
20:19-20, *53, 133*
21:10–34:12, *131*
22:6-7, *52, 61*
22:7, *56, 59*
22:9, *123*
23:25, *131*
24:19, *76*

24:20, *76*
24:21-22, *77*
25:4, *35*
25:5-10, *72, 73*
26:1-11, *20*
26:12-15, *18*
28, *16*
28:51, *131*
29:8, *121*
29:28, *16*
30:9, *109*
30:18, *24*
30:19, *16*
32:47, *24*
33, *121*
33:28, *131*

Joshua
7:14-15, *135*
12–24, *121*
18:1, *119, 120*

Judges
11:27, *140*

Ruth
1:5, *74*
1:8, *74*
1:11-13, *74*
1:16-18, *75*
2:1, *75*
3:11, *75*
4:11, *75*
4:16-22, *75*

1 Samuel
8:14, *76*

2 Samuel
8:11, *119, 120*

1 Kings
4:25, *62*
5:4, *119*

5:11, *76*
5:25, *76*

2 Kings
3:19, *61*
18:31, *62*

1 Chronicles
22:18, *119*
27:28, *77*

Job
24:6-10, *71*
31, *71*
31:16-22, *72*
31:38-40, *23*
38:12, *49*
38:16, *49*
38:39-40, *49*
39:1-2, *49*
39:5-6, *50*
39:26-27, *49*

Psalms
8, *10*
8:3-9, *10*
8:9, *49*
104, *50*
104:10-11, *50*
104:16-18, *50*

Proverbs
14:4, *121*

Isaiah
1:17, *67, 89*
5, *98*
5:8, *121*
13:2-13, *96*
13:6, *98*
13:9, *98*

Jeremiah
46:10, *98*

Scripture Index

Ezekiel
7:10, *98*
13:5, *98*
30:3, *98*
34:3-4, *127*
47:1-12, *103*

Daniel
2:31-35, *98*
11:39, *119*

Hosea
12:12, *77*

Joel
1:15, *98*
2:1, *98*
2:11, *98*
2:30-32, *98*
3:4, *98*
3:14, *98*

Amos
5:18, *98*
5:20, *98*

Obadiah
15, *98*

Zephaniah
1:7, *98*
1:14, *98*

Malachi
4:1, *98*
4:5, *98*

New Testament

Matthew
1:1, *75, 101*
21:33-46, *98*
24:35-44, *98*
24:37, *100*
25:32-33, *33*

Mark
1:13, *140*

Acts
2:20, *98*
2:20-21, *98*

Romans
1–8, *140*
3:23, *101*
5:12-21, *102*
8, *100, 102, 113*
8:18-25, *101*
8:20, *12*
8:21, *102, 109*
8:23, *102*
12:2, *110, 112*

1 Corinthians
4:3, *140*
5:5, *98*
15:23, *98*
15:42-58, *101*

2 Corinthians
5:17-21, *14*

Colossians
1:16, *104, 109*

1 Thessalonians
3:13, *98*
5:2, *98*
5:2-3, *94*

2 Thessalonians
2:1-3, *98*
2:2, *98*
2:7-8, *98*

James
1:27, *67*
5:8, *98*

2 Peter
3:10, *98*
3:10-13, *93*

Revelation
6:12-14, *94*
6:17, *94*
21:1, *68, 94*
21:1-2, *103*
21:11, *103*
21:18, *103*
21:23, *103*
21:25, *103*
22:1, *103*
22:1-2, *103*
22:3-5, *103*